from
Black Clouds
to
Black Holes
Third Edition

WORLD SCIENTIFIC SERIES IN ASTRONOMY AND ASTROPHYSICS

Editor: Jayant V. Narlikar
Inter-University Centre for Astronomy and Astrophysics, Pune, India

---

*Publication cancelled.

World Scientific Series in Astronomy and Astrophysics – Vol. 13

from Black Clouds
to Black Holes

Third Edition

Jayant V. Narlikar

Inter University Centre for Astronomy and Astrophysics, India

World Scientific

NEW JERSEY · LONDON · SINGAPORE · BEIJING · SHANGHAI · HONG KONG · TAIPEI · CHENNAI

Published by

World Scientific Publishing Co. Pte. Ltd.

5 Toh Tuck Link, Singapore 596224

USA office: 27 Warren Street, Suite 401-402, Hackensack, NJ 07601

UK office: 57 Shelton Street, Covent Garden, London WC2H 9HE

British Library Cataloguing-in-Publication Data
A catalogue record for this book is available from the British Library.

World Scientific Series in Astronomy and Astrophysics — Vol. 13
FROM BLACK CLOUDS TO BLACK HOLES
Third Edition
Copyright © 2012 by World Scientific Publishing Co. Pte. Ltd.

ISBN-13 978-981-4350-37-2
ISBN-10 981-4350-37-0

Printed in Singapore by Mainland Press Pte Ltd.

Preface

I am happy to present this revised edition of my book *From Black Clouds to Black Holes*. This book updates the subject of stellar evolution and includes new inputs, such as extrasolar planets, variable stars and binaries etc. In its overall style and format this book is similar to the earlier edition.

I thank Professor K.K. Phua for encouraging me to write this revised version and hope that the readers will find it useful to have it on their desk. I thank my colleague, Vyankatesh Samak for preparing the typescript of this manuscript. I am also grateful to Arvind Paranjpye for help in preparing the diagrams.

Jayant V. Narlikar
IUCAA, Pune, India

Preface to the Second Edition

Man's age-old curiosity about the heavens has found expression in a variety of ways ranging from poetic fancies to profound philosophical thoughts and deep-seated religious beliefs. Science entered into the picture comparatively recently, but it has done so with commendable success.

Indeed, it seems audacious even to attempt an application of the scientific laws, discovered under very limited conditions here on the Earth, to heavenly bodies that lie light years away and even beyond. Unlike his counterpart in other branches of science who can handle or control the systems under his study in the laboratory, the astrophysicist is forced to watch them helplessly from great distances. It is, therefore, all the more creditable that he should have achieved any measure of success at all.

In fact, this success has generated confidence in the wide applicability of the laws of science discovered here on the Earth. One can go even further and assert that astronomy has provided fresh inputs into our understanding of the physical world that would not otherwise have come at all. For example, but for the astronomical observations of the planets and satellites, the law of gravitation (which is one of the four fundamental laws of physics) would have long remained undiscovered.

No branch of astrophysics can claim as impressive a record of successes as the study of the internal constitution of stars and of how stars evolve with time. Long before laboratory physicists could even think of nuclear fusion as a source of energy, the astrophysicists had stellar models which worked on this process. It is through the study of stellar evolution that we receive a clue to the origin of the variety of different chemical elements found in the Universe. When following through the life story of stars, we discover that even such esoteric objects as black holes have a natural place in it.

The aim of this small book is to tell that story in as nontechnical a language as possible, and to convey to the lay reader the excitement that drives the astrophysicist to even more ambitious tasks that lie ahead. For example, the mystery of how the galaxies consisting of hundreds of billions of stars came into existence, the riddle as to what source of energy powers the highly luminous but highly compact quasars, and of course the most fundamental question of the origin of the Universe as a whole constitute some of the challenging fields in astronomy. It is because of his ability to understand what is going on inside stars that the astronomer feels confident today that these questions also will receive at least partial answers in the not-too-distant future.

The first four chapters in the book prepare the groundwork for the stellar odyssey. The story of star's life begins in Chapter 5, in the visually dark interstellar clouds. It ends in Chapter 10, when a star unwise enough to retain too much weight late in its life ends in a black hole. The intervening chapters describe how the changing physical characteristics of a star can be explained with the physics that we know here and now. But, as outlined in the final 11th chapter of the book, there are still some mysteries left even in this relatively well understood part of astrophysics.

While attempts have been made to keep the description as nontechnical as possible, a few simple formulae could not be avoided. Likewise, occasional use of powers of ten and the logarithmic notation had to be made; it is hoped that the uninitiated may find the Appendix at the end helpful in this context.

Ten years have elapsed since the first edition of the book came out. In the meantime the subject has moved on and a new updated edition became desirable to include some of the recent developments. This revised edition includes, therefore, a few additions while keeping the original format undisturbed.

The book owed its origin to an invitation from Professor K.K. Phua to write for the World Scientific Publishing Company, and I am grateful to him for his help and encouragement for the original and the revised editions. I also thank Mr. Santosh Khadilkar for typing the manuscript and Mr. Arvind Paranjpye for help in preparing the figures.

Pune, India   Jayant Narlikar

Contents

Prologue

As the bright star Alpha Centauri faded into the background the three occupants of the tiny spaceship woke up from their mechanically induced sleep. Time for action was fast approaching.

The Professor was the first to come back to normal. But if his two pupils and assistants were still feeling groggy when he summoned them, it was not long before his crisp instructions alerted them to their imminent responsibilities. Reminding them of the basic purpose of their mission, he added:

"Very soon we will be close to the planet X, that is known to host intelligent and relatively well-developed forms of life. At a suitable stage I will beam you both to the planet's surface. Of course, you will be suitably equipped for survival there as long as you wish. Thanks to your advanced equipment, your presence on X will go unnoticed by its inhabitants. Your primary assignment is to find out about these creatures, how they are born, how they grow old, how they live and how they die. Keep in touch with me through your signaling device: when you have finished, I will beam you back." The Professor waited for any questions.

"Shall we both move together?" asked Sunya, who was hardworking but somewhat simple-minded.

"No! It is better if you investigate independently. I will put you down in different parts of the planet."

"Sir! Is it known if X has a large population of such creatures?" the query came from Sunya's somewhat sleepy-looking companion.

"Yes! But, Purna, why did you ask?"

"It will simplify my task in that case," said Purna cryptically. The Professor understood and smiled. He knew Purna was lazy but very intelligent.

Sunya did not see the point. A large population, he thought, would make their job more difficult. But before he could seek clarification it was time for them to be beamed out.

It was a long time before Sunya completed his survey and was beamed back. He had worked hard and felt sure that he would be the first to return. To his surprise, he found Purna already back and the Professor waiting impatiently.

"You took a long time! Hope your report is a good one. Tell me what did you do and what did you find." The Professor wasted no time.

"Sir! The planet is called Earth and the inhabitants are called human beings. To make a thorough study, I decided to observe one member of these species all through its life. I saw it right from its birth, watched it grow into an adult, then live on till it got old and finally died. I have recorded all details of its life right here." Sunya produced a small piece of equipment and continued. "Naturally, the whole procedure took a long time — about seventy-six revolution periods of the Earth about its star, the Sun." Sunya was proud of the effort he had put into this job and was expecting to be applauded.

But the Professor was not pleased. "Yes, I know you spent seventy-six years to get the details of one human being. But what do you know about the variety of humans on Earth? Are you sure that the member you observed is a typical member of the human population or at least a part of it?"

"Sir, there was hardly time available for me to find answers to these questions — for the human population is so large!" Sunya was aggrieved. "May I know how Purna fared in his task?"

"It may interest you to know that he came back within a year and has brought back much more valuable information than what you seem to have collected." The Professor handed him the report submitted by his rival.

As Sunya read on, understanding began to appear on his face.

Chapter 1

On Stars and Humans

At this stage we say *au revoir* to the occupants of the spaceship but elaborate a little on how Purna had gone about his mission.

The clue to his approach lies in his remark that his task was simplified by the largeness of the human population. For, instead of watching just one human being, he surveyed human groups in largish habitations like cities in different countries. In a typical city he saw human beings at different stages of life — from pre-birth stage right through to old age and death. Besides, by observing their physical characteristics like height, weight, colour of hair, texture of skin etc., Purna was able to form a fairly accurate view of how these characteristics of a typical human being change with age.

For example, Figure 1.1 illustrates the distribution of height and weight amongst human beings in a typical city. Notice that there is a trend in these points which rises from low values of height and weight and reaches a plateau wherein there is not much variation in height but a larger variation in weight. Also there are many more points on the plateau than in the rising part of the trend. What does this distribution imply?

First we note that the distribution of points in Figure 1.1 relates to one point of time: it gives the characteristics of the human population at the time of the survey. All the same it contains clues as to how an individual human being *changes* its characteristics *with time*. For, a child at birth is at the left-hand end A of the distribution where both

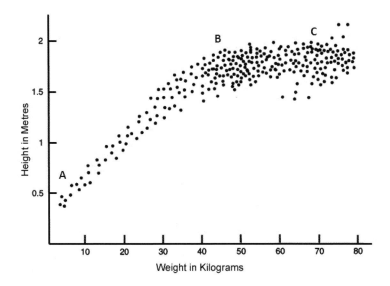

Figure 1.1: A schematic plot of height and weight of a group of human beings residing in a city.

the height and weight are small. As he or she grows in age, these values increase so that the person "moves up" the distribution until reaching a full adult height at point B. Thereafter the height will not increase but the weight might well do so! The section BC thus contains the bulk of the adult population reflecting its relatively steady state. The fact that there are more points on the section BC than on the section AB tells us that on an average the human being spends a greater fraction of life as an adult than as a child growing to adulthood.

The growth in age of a human being can also be correlated with the greying of the hair on the head and also, in most cases, with the thickness of hair growth. Likewise age can be correlated with the texture of the skin which evolves from a smooth texture to a wrinkled condition.

It is evident that by collecting such information about the human population, Purna brought considerably greater knowledge about the Earthmen than Sunya could by his extensive observation of just one human being. The method adopted by Purna had the added advantage that it took much less of his time.

⋆   ⋆   ⋆

This book is not about humans; it is about stars. Our mission is to try and put together a credible account of how stars evolve through their lives.

However, the above example guides us as to what approach we, as human observers, should use to find out how stars change as they get older: how they are born and how they die. Fortunately, as Purna observed in the case of humans, we have a large stellar population to survey. If we are clever about it, we should be able to observe the different physical characteristics of stars in a group and use those observations to tell us how these characteristics change in a star as it ages.

The alternative to this approach, the method used by Sunya is clearly not going to help us here. For example, for us the Sun is the nearest star and following Sunya's method we may be tempted to follow its evolution all through its life. The Sun, however, does not perceptibly change during a typical human life span; in fact, it has not so changed during the entire recorded history of human civilization. Indeed, as we shall find later, the characteristic time scale for change in a star like the Sun is thousands of millions of years!

Returning therefore to the method used by Purna we should first find out what are the relevant physical characteristics of stars that can be observed from here and whose changes are correlated with the star's age. In the earlier chapters we will give a discussion of these properties.

Armed with this information we will then be able to follow the remarkable story of a star's life, a story that begins in a black interstellar cloud and which may well end with the star becoming a black hole.

Chapter 2

Light: The Storehouse of Information

A casual glance at the starry sky on a clear night may give the false impression that all stars are alike. This impression is soon corrected when we look at individual stars more carefully. We then find that some stars are brighter than others. We may even be able to notice subtleties of colour: the overall golden appearance may have tinges of red in some cases and even of blue in some others. We might also be able to notice the differences in size: some stars appear bigger than others.

The professional astronomer, of course, no longer relies on his eyes as the ultimate source of information. It was Galileo Galilei who in 1609 used the telescope as an implement to aid his eyes for making astronomical observations: he thus became the first astronomer to use a telescope (Figure 2.1). With the help of the telescope Galileo was able to see things, like the satellites of the planet Jupiter, or phases of planet Venus, even dark spots on the solar disc which he could not see with naked eyes.

The telescope used by Galileo was small by modern standards, but it heralded a revolution in astronomical observing. The realization that instruments like a telescope can improve upon our naked eye perception of the heavens led to more and more efficient instrumentation. To show its appreciation of Galileo's use of the telescope, the world astronomical community celebrated the year 2009 as the International Year of

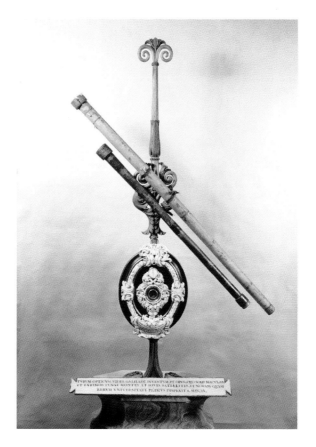

Figure 2.1: Galileo's telescope which brought in a revolution in observational astronomy.

Astronomy. (The year 2009 marked the end of the fourth Centenary of Galileo's first use of the telescope.)

Galileo's example set the pattern of progress of astronomy from his time to the present, a pattern that will continue into the future too. The underlying theme of this progress is the following. Our main source of information about the Universe is the light that comes from its far corners: the progress we make is in our ability to collect and interpret that information. Just as Galileo was able to discover new objects in the Universe with his telescope so do astronomers of today achieve greater

Figure 2.2: James Clerk Maxwell (1831–1879).

perception of the heavens than their predecessors could. The key to this perception lies, in the last analysis, in the information carried by the light we receive from "up there".

So, before we talk about stars let us examine first the nature of light itself. For, only then will we be able to appreciate the information conveyed by starlight.

Light is a wave

It was in the 1860's that physicists were able to make a breakthrough in understanding the nature of light. The work of James Clerk Maxwell (Figure 2.2) demonstrated theoretically that what we call light is in fact an *electromagnetic wave*. What is an electromagnetic wave?

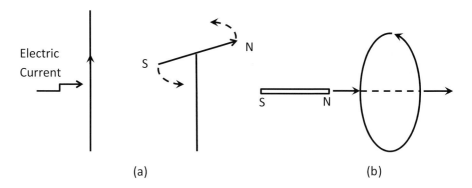

Figure 2.3: (a) The magnetic compass needle gets deflected when electric current flows in a wire in its vicinity, thus demonstrating that the flow of the current generates a magnetic force. This effect was discovered by Andre-Marie Ampere (1775–1836). (b) An electric current flows in the closed loop of wire when a magnet is swiftly moved through it. It was Michael Faraday (1791–1867) who first demonstrated that a changing magnetic field through an electric circuit generates electric currents in it.

The life of high technology today brings us in contact with many effects of electricity and magnetism. Although initially, the phenomena of electricity and magnetism were discovered separately, it gradually became clear that they are inter-related. Figure 2.3 describes two experiments that demonstrate this interrelationship. The electric motor and the dynamo which are so essential for today's technology make use of electromagnetic effects like these. Basically, we find that when electric fields change with time they produce magnetic fields and vice-versa. But it was Maxwell whose equations brought together all the different properties of electricity and magnetism in a simple unified form. And out of those equations emerged the above mentioned result that light is an electromagnetic wave.

Figure 2.4 describes the simplest electromagnetic wave, and to understand its nature we can be guided by the analogy of ripples created on a still pond by throwing a pebble into it. The ripples appear to move outwards from the spot where the pebble hits the water surface. But this motion is illusory in the sense that there is no outward physical motion of water particles. These particles simply move up and down in

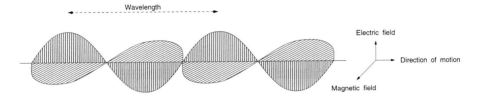

Figure 2.4: The wave indicates ups and downs in the strength of the electric disturbance. The magnetic disturbance behaves similarly. As the wave advances to the right these disturbances rise and fall periodically.

the same place, but in such a coordinated way that they produce the appearance of an outward-moving wave.

Likewise the electromagnetic wave in Figure 2.4 is made out of changing electric and magnetic disturbances. The two disturbances take place in perpendicular directions. The arrows indicate, through their lengths, the strengths of these disturbances, called in technical jargon "fields". The undulating shape of these fields is indicated in the figure.

The distance between two successive points of positive maximum (crest of the wave) is called the *wavelength*. It is usually denoted by λ.

What is the time that has elapsed between two peaks in succession? Clearly during this period the electric field and the magnetic field have gone through one cycle of changes, from maximum positive through maximum negative back to maximum positive. The number of such cycles in a unit of time (usually taken as one second) is called the *frequency* of the wave. It is often denoted by the symbol ν.

By examining the shift of the wave from peak to peak in Figure 2.4 we see that in unit time there will be ν cycles and in each cycle the wave will move a distance λ. Therefore the *speed* of the wave, being the distance travelled by it in one unit of time is given by

$$c = \nu \times \lambda .$$

Returning to Maxwell's equations, we now state the important result to emerge from them: that all electromagnetic waves travelling in vacuum do so with a speed equal to the speed of light in vacuum; in fact *light itself is a manifestation of electromagnetic waves*; when a light ray passes through space it generates ripples of electric and magnetic fields along its path precisely in the form of a wave.

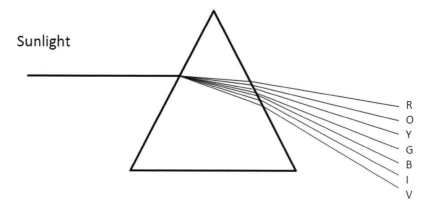

Figure 2.5: Sunlight passing through a prism splits into a spectrum of colours ranging from red to violet. Colour names are indicated by their initial letters.

The speed of light in vacuum is known very accurately today:

$$c = 2.9979 \times 10^8 \text{ metres per second.}$$

What are the typical values of ν and λ?

Not all light waves have the same values of ν and λ. Indeed, we can do a simple experiment which Isaac Newton had done in his studies of light. This experiment consists of holding a glass prism in sunlight so that as the sunlight passes through the prism [*see* Figure 2.5], we notice that it is split into different colours. This splitting into a band of colours, known as the *spectrum*, happens because a light wave gets bent at the prism surface (a phenomenon known as *refraction* of light) and the amount of bending depends on the wavelength of the wave. Sunlight happens to contain waves of different wavelengths. We see from Figure 2.5 that the red coloured light is bent the *least* and the violet coloured light is bent the *most*. Maxwell's theory tells us that bending decreases as wavelength increases. Consequently we arrive at the important conclusion that red light has the longest wavelength and violet light the shortest.

Table 2.1 below gives the approximate ranges of wavelength corresponding to the different colours of the rainbow. These wavelengths are usually expressed in units of Angstroms (Å = ten billionth part of a metre) or nanometres (1 nm = a billionth part of a metre).

Table 2.1: Colours and wavelengths.

Colour	Wavelength Range (in Å)
Violet	3900-4300
Indigo	4300-4500
Blue	4500-4900
Green	4900-5770
Yellow	5770-5970
Orange	5970-6220
Red	6220-7700

Notice that the wavelength range is limited to 4000Å–8000Å (400nm–800nm) approximately. What Table 2.1 therefore tells us is that the light we see with our eyes covers a very limited range of wavelengths: there must be other forms of light in nature that we *cannot* see. These forms would have wavelengths shorter than 4000Å or longer than 8000Å. Indeed experiments have revealed the existence of several "invisible" light forms, ranging from the "radio waves" at the very longest end of λ to the "gamma rays" at the very shortest end. The approximate wavelength and frequency ranges of these forms are given in Table 2.2. (Note that Å is *not* an appropriate unit to describe all the entries of λ in this table!)

Evidently we cannot expect to get a comprehensive knowledge of the universe simply from visible light, which, according to the above table forms only a tiny window. Any information receivable through other wavelength windows cannot be ignored. We will return to this point when we consider starlight as a source of information.

We now turn our attention to another property of light which at first sight appears to contradict its wave nature.

Light as a collection of particles

Even as physicists were telling themselves in the 1860's that they had now understood all about light, additional information was coming in, mainly from astronomy, that did not quite fit into the above picture.

Table 2.2: Different types of electromagnetic waves.

Wave Type	Wavelength Range (metres)	Frequency Range (cycles per second)
Radio waves	longer than 10^{-1}	less than 3×10^9
Microwaves	$10^{-3} - 10^{-1}$	$3 \times 10^9 - 3 \times 10^{11}$
Infrared light	$8 \times 10^{-7} - 10^{-3}$	$3 \times 10^{11} - 3.75 \times 10^{14}$
Visible light	$4 \times 10^{-7} - 8 \times 10^{-7}$	$3.75 \times 10^{14} - 7.5 \times 10^{14}$
Ultraviolet light	$10^{-8} - 4 \times 10^{-7}$	$7.5 \times 10^{14} - 3 \times 10^{16}$
X-rays	$10^{-11} - 10^{-8}$	$3 \times 10^{16} - 3 \times 10^{19}$
Gamma rays	shorter than 10^{-11}	above 3×10^{19}

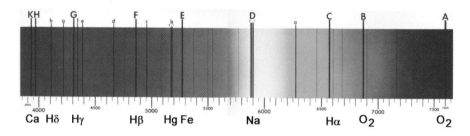

Figure 2.6: The dark Fraunhofer lines shown against the background of the continuous solar spectrum. The chemical symbols of atoms responsible are given underneath.

Figure 2.6 shows a spectrum of sunlight taken with a more sophisticated instrument than a simple prism. Apart from the rainbow of colours there are some dark lines in the spectrum. These lines were first found in the solar spectrum in 1814 by J. Fraunhofer and even after the "full" understanding of light with Maxwell's equations, it was not possible to see why such sharp dark lines should appear in an otherwise continuous range of colours.

Laboratory experiments in spectroscopy yielded a different class of spectral lines as shown, for example, in Figure 2.7. In contrast to the dark lines found by the astronomers in the spectra of the Sun or other stars, these were bright lines that stood out against the background

Figure 2.7: The bright spectral lines arising from the excited atoms of sodium in vapourized form. These are called the D-lines.

continuum spectrum. One could say, on empirical grounds, that the dark lines of Figure 2.6, known as *absorption lines*, arose from absorption of light while the bright lines, called *emission lines*, arose from the emission of light. But the lines correspond to very narrow ranges of wavelengths — not more than a few Angstroms — and it was hard to see how a material could selectively absorb or selectively emit light in such narrow ranges of wavelengths.

The mystery was resolved in the twentieth century with the realization that light has a dual nature. Apart from being distributed like waves, it can also be looked upon as being made of tiny packets of energy called *photons*. A photon of light of frequency ν carries an energy $E = h \times \nu$, where the constant h is a universal constant, called *Planck's constant*. The idea that light could be so described had in fact first been suggested by Max Planck in the year 1900 and it marked the beginning of the so-called *quantum theory*. But Planck's reasons for making this assumption were different and we will come to them later.

Quantum theory explains the occurrence of emission lines and absorption lines in the following way. Let us take the example of a gas made up of the simplest atom, that of hydrogen. Figure 2.8 describes the quantum picture of this atom. It has a positively charged heavy particle called the *proton* in its nucleus and a negatively charged light particle called the *electron* going round it. Before quantum theory came into existence this motion of the electron around the central proton was a mystery. For, Maxwell's equations led to the conclusion that an electron circling around must inevitably radiate electromagnetic waves. The energy carried by these waves must come from the electron's own energy of motion. Because of this energy loss the electron's orbit steadily shrinks and it eventually winds up into the central proton. And, the most disturbing aspect of this conclusion was that the time scale during which all this happens is extremely short, of the order of 10^{-23} second!

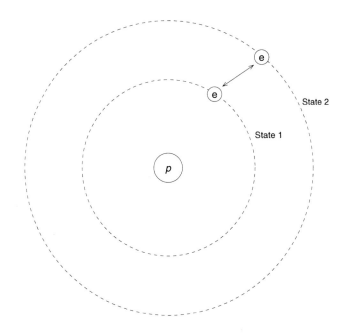

Figure 2.8: The dotted orbits around the central nucleus schematically describe the different states of energy of the electron in a hydrogen atom. The energies increase outwards. When the electron jumps from an outer orbit to an inner one it radiates energy. Energy needs to be supplied for the reverse jump.

So the mystery was, how does the hydrogen atom manage to maintain its structure for a long time. What is it that prevents the electron from falling into the proton? The methods of Newton and Maxwell failed to provide an answer. However, the new rules of motion according to quantum mechanics provided a deeper understanding. These rules introduced the concept of discreteness into the traditional picture of the atom. According to quantum theory the electron is allowed to have only a discrete set of values for its energy and consequently it selects one of a discrete set of orbits compatible with its energy. The smallest orbit in this set is one in which the electron has the least energy. The orbit of the electron cannot become any shorter than this smallest orbit and so the electron cannot ever fall into the proton. If it is supplied energy from outside, however, it will jump to a larger orbit, provided the energy supplied is of the right amount.

What is the "right" amount of energy? In Figure 2.8 we see two typical permitted orbits representing states 1 and 2, in which the electron has total energies E_1 and E_2, E_2 being greater than E_1. To knock the electron into the orbit of energy E_2 from its original orbit of energy E_1 the amount of energy that must be supplied is $E_2 - E_1$.

Although Figure 2.8 is schematic it conveys the important point that these orbits form a discrete set, with the corresponding electron energies increasing outwards. It explains why the electron has to "jump" from one orbit to another rather than make the transition in a smooth manner.

Let us now bring into this picture the light wave as the agent providing energy. In our example of Figure 2.8 we need light to provide energy $E_2 - E_1$ in order to knock the electron from energy level E_1 up to level E_2. According to the quantum theory of light, only a photon of a specific frequency can achieve this knocking off procedure. This frequency is given by the rule stated earlier:

$$\nu = \frac{E_2 - E_1}{h}.$$

Thus we arrive at a picture of how radiation composed of photons is absorbed selectively at a discrete set of frequencies by intervening atoms. If the photons are of the right frequency they knock the electrons of the atoms up the energy ladder. And, the photon getting absorbed in this process reduces the intensity of radiation *of that specific frequency*. Thus the dark lines in the solar spectrum arise from the absorption of specific frequencies of radiation by intervening atoms mostly in the Sun's atmosphere.

For example, the line identified at a wavelength of 6563Å and designated as the (C) line by Fraunhofer is seen to arise when the electron in a hydrogen atom makes a jump from the second to the third energy level. Quantum theoretical calculations of the various possible energy levels in different atoms have been made and theoreticians have lists of frequencies and wavelengths of spectral lines associated with the electron jumps in these atoms. An experienced astronomer can thus identify a dark line of specified wavelength in a star's spectrum with the absorbing atom responsible. The prominent Fraunhofer lines are thus known to be due to absorption by atoms of hydrogen, sodium and calcium in the Sun's atmosphere.

The bright lines, or the emission lines arise from the reverse process. When an electron in energy level E_2 jumps *down* to energy level E_1, it emits a photon of frequency $\nu = (E_2 - E_1)/h$. This jumping down (unlike the jumping up) can happen spontaneously without the need for radiation to be present. If radiation of the "right" frequency is present that also helps in the process of the downward jump. Emission (or absorption) so assisted by ambient radiation is called *stimulated* emission (or absorption). Unassisted jump down is called *spontaneous* emission.

Thus we arrive at the important result that if a certain atom acting as an absorber produces dark lines of specific frequency in a light spectrum, then that atom acting as an emitter will produce bright emission lines *at the same frequency*.

In a star, as we shall see later, the hot external surface in some cases makes the electrons in the atom acquire high energies. These electrons jump down the energy ladder and produce emission lines in the star's spectrum. If we can identify those lines we know what atoms are present on the external surface of the star. Likewise radiation coming from the hotter inner regions of the star produces absorption lines as it passes through the relatively cooler stellar atmosphere.

From dark and bright spectral lines we now turn our attention to the continuous distribution of light over waves of all frequencies. In particular, we will look at a distribution that has a special status in fundamental physics and which also happens to be highly relevant to the type of spectrum that is received from a star.

Black body radiation

Imagine the interior of an oven which is being heated. Initially some parts of the oven will be hotter than others. These hot parts radiate heat towards the cooler ones whose temperature begins to rise. This process of give and take goes on and on until all points inside the oven are "equally hot": there is no net flow of heat from one point to another. In an ideal oven (whose walls do not lose any heat to the surroundings) the above equilibrium situation is soon attained.

Now "heat" inside the oven is nothing but the electromagnetic waves knocking about between points. In an oven heated to say 250°C, the waves are predominantly in the infrared region. If the oven were heated to 5000°C the waves would be predominantly in the visual range.

Here we take up another example. Suppose we start heating a metal rod. It first becomes "red hot", that is, its colour changes to a reddish glow. If heated further the colour changes: a piece of "white hot" metal is hotter than one which is "red hot". If we compare these examples of the oven and the iron rod, with Tables 2.1 we notice that the predominant wavelength of radiation emitted by the heated body is related, through its colour, to its temperature.

Our idealized oven is "black body". Since no radiation is lost from the walls, the body is "black" to an outside observer! It is internally hot, however, and its interior has reached a state of equilibrium when each point is absorbing as much radiation as it is giving out. Max Planck introduced new ideas from quantum theory to calculate how much energy is carried by waves of different frequencies.

Figure 2.9 describes these intensity distributions for black bodies of different temperatures. All curves share one property. The distribution of intensity is not uniform for all frequencies: the energy carried by very small or very large frequencies is relatively small, with the peak of emission occurring somewhere in the middle. If we compare all the different curves of Figure 2.9 we find that the frequency at which the intensity peaks increases with temperature. This result, discovered by Wien in 1896 on empirical grounds is known as *Wien's law*. The scale of temperature used in Figure 2.9 is called the *absolute scale*.

What is absolute temperature? It is temperature measured on a scale that describes quite naturally the physical meaning of the property of temperature. Temperature is a measure of the internal microscopic motions going on within a body, the motions of its constituent atoms or molecules. The centigrade scale (or the Celsius scale) used in everyday life starts with 0°C at the freezing point of water and reaches 100°C at the boiling point. But even in ice, cold as it may be, there are internal motions going on. So the Celsius zero does not convey the natural state of affairs within the body. *The absolute scale does.* The zero on the absolute scale for a substance really means absolutely no internal motions

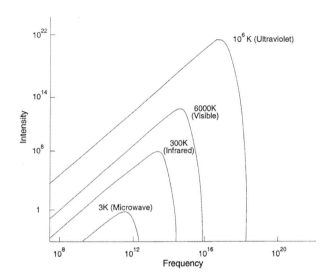

Figure 2.9: Black body intensity curves for different temperatures. The type of radiation at peak frequencies is shown. Notice that the peak frequencies increase with temperature. (Both intensities and frequencies are on logarithmic scale.)

of *any* kind in that substance! The absolute zero is approximately equal to $-273°$C, while one unit of this scale, 1 K, equals $1°$C. Thus on the absolute scale, water freezes at the temperature of 273 K and boils at 373 K. The letter K stands for Kelvin. Lord Kelvin (1824–1907) did pioneering work in the subject of *thermodynamics*, a subject that relates the nature and properties of heat to the mechanical motions of the microscopic constituents of physical systems.

Taking all this into account we see the following connection so far as the black body radiation is concerned:

Temperature → Frequency → Predominant type of radiation.

A black body of temperature 3 K will predominantly radiate microwaves of frequency 3×10^{11} Hz. A black body of temperature 6000 K will predominantly radiate green light in the visual range while a black body heated to 10^6 K will radiate mainly X-rays, ... and so on! By measuring the peak wavelength of black body radiation we can therefore determine its temperature.

Of course, this result in its idealized form is useless for an outside observer who cannot see the black body. But we can make it useful by allowing small concessions. If the black body is punctured with tiny holes through which a very small quantity of radiation escapes then we can use that radiation as a probe of what is present within the body. Thus by examining the hot radiation leaking out of an imperfect oven we can estimate its internal temperature. As we shall see in the next chapter, this property enables us to determine the temperatures at the surfaces of stars because they happen to be like an imperfect oven.

Chapter 3

The Vital Statistics of Stars

The last chapter indicated how light from a distant and physically inaccessible source can tell us, in many different ways, about some interesting physical characteristics of that source. In Chapter 1 we saw how the physical characteristics of a human population vary from member to member and this variation tells us something about the life of a human being. In our present task of putting together a star's biography, we will next see how, in a stellar population, the important characteristics of stars as conveyed by their light vary from one star to another. The key role in setting out these vital statistics of stars is played by a diagram which is somewhat analogous to the height-weight diagram for human beings shown in Figure 1.1. This diagram was first introduced into astronomy by Ejnar Hertzsprung (in 1911) and Henry Norris Russell (in 1913) for different populations of stars. It is commonly known as the *Hertzsprung-Russell diagram* or simply the *H-R diagram*. We will refer to it by the shorter name. To understand what the H-R diagram is about we have to introduce some technical jargon.

Magnitudes

There are two ways of comparing different amounts of a given substance. For example, businessmen talk of profit and loss which are basically the difference in the selling price and buying price. The difference in this case is expressed through the *subtraction* of one quantity from the other. The second way in which a difference can be expressed is through *division*. For example, if in a war the opposing armies had strengths 10,000 and 40,000 respectively we say that one side was numerically four times stronger than the other. The essential difference in the fighting power here is conveyed not by subtraction but by the division of 40,000 by 10,000.

When we compare the brightness of two luminous objects we again use division to express the difference. Thus, given two electric bulbs of powers 1000 watts and 10 watts respectively we do not say that the first bulb is brighter than the second by 990 watts, since the subtraction process does not convey the perceptible difference in the powers of the two light bulbs. Rather, we say that the first bulb is hundred times stronger than the other — the factor 100 derived from dividing 1000 by 10. Let us now make a transition from light sources of human creation to the large cosmic ones.

What applies to light bulbs also applies to stars which are sources of light too. The power (that is, the amount of energy radiated per second) by a star is called its *luminosity*, and it can also be measured in watts. Only that the number of watts measuring the luminosity of a typical star will be much larger than numbers like 10 or 1000 which we used above for light bulbs. For example, the luminosity of the Sun is 4×10^{26} watts (four hundred million million million million watts).

Clearly we cannot compare the powers of stars with those of light bulbs. But we can compare them with one another. Astronomers wanted to evolve something like a ranking-scheme wherein they could rank the brightest star as number 1, the next brightest as number 2 and so on. Such a scheme would, however, be qualitative in character since it cannot tell us how bright star number 1 is compared to star number 2.

To quantify such a ranking the astronomers have devised the *absolute magnitude scale* for luminosity. The scale has ranks $1, 2, 3, \ldots$ such that a star of rank 1 is as many times brighter than a star of rank 2 as the star of rank 2 is compared to a star of rank 3 ... and so on. These ranks are related to the so-called absolute magnitudes.

The factor by which the luminosity decreases as we ascend the magnitude scale is approximately 2.512. What is so special about this number? It is such that multiplied to itself five times it gives the factor of

100. In other words, a star of magnitude 1 will be 100 times *brighter* than a star of magnitude 6.

Of course, so far we have not mentioned which star is to have the absolute magnitude 1. In order that we can compare the luminosities of different stars we must observe them all from the same distance. In practice we clearly cannot do this since we are constrained to observe the stars from the Earth and, as we shall see in the following chapter, the stars lie at varying distances from us.

To get round this difficulty astronomers have to introduce another magnitude scale which is more realistic in the sense that it includes the effect of distance also. To distinguish it from the *absolute* magnitude scale, this new scale is called the *apparent* magnitude scale. Let us look at this scale first before we try to answer the question "What luminosity would entitle a star to have the absolute magnitude 1?"

The relation between the absolute and apparent magnitudes can be understood with the help of the so-called "inverse square law of illumination". Let us find out what this law is about with the help of our two light bulbs, the strong one of 1000 watts and the weak one of 10 watts.

Suppose we observe the two bulbs from the same distance. It is obvious that the 1000-watt bulb will look much brighter than the 10-watt one. Now let the former be moved further and further away from us. It will begin to appear less and less bright. At a certain distance its brightness will have become so diminished that it will appear as faint as the 10-watt bulb which is nearby. According to the *inverse square law of illumination* this distance, at which the 1000-watt bulb looks as bright as the nearby 10-watt bulb, is ten times the distance of the latter bulb. For, the brightness of an object as seen by an observer diminishes in inverse proportion to the square of its distance from the observer. By moving it 10 times further away we have diminished the apparent brightness of the 1000-watt bulb by a factor $10 \times 10 = 100$ so that it now matches the brightness of the 10-watt bulb (see Figure 3.1).

Why this law arises can be explained with the help of Figure 3.2. We have a source of light O which illuminates all directions equally. Σ is a spherical surface surrounding O with O at its centre. Euclid's geometry tells us that if the radius of Σ is d, its total surface area is $4\pi d^2$. As seen

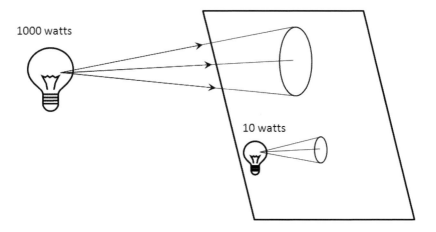

Figure 3.1: The small patch on the screen illuminated by a 10-watt bulb is as bright as the large patch illuminated by the 1000-watt bulb ten times farther away from the screen. (Figure not drawn to scale.)

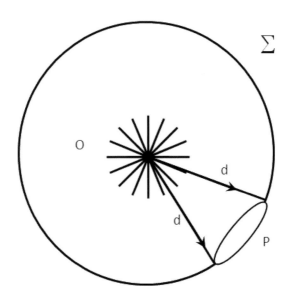

Figure 3.2: The light from O is uniformly distributed over the surface of the sphere Σ of radius d. A typical point P on the sphere receives a proportionate fraction of this light across a unit area of Σ around P.

in Figure 3.2, all light from O is distributed uniformly across this area. If L is the luminosity of O, the amount of its energy falling on unit area of Σ per second, is obtained simply by dividing L by the total area of Σ:

$$l = \frac{L}{4\pi d^2}.$$

l is the illumination that any point P on the surface Σ measures as coming from O. If P moves further away, d increases and l decreases in inverse proportion to the square of d. It is l therefore that conveys the apparent brightness of the source of light, as seen by a typical observer P.

Now imagine O to be a star and P an observer like us. What P measures is not L but l. While looking at different stars P is comparing not their L-values but their l-values. Not knowing their distances P has only these l values which indicate how much brighter one star *appears* compared to another from P's particular location.

The *apparent magnitude* is the scale which ranks stars on their l-values in the same way that the absolute magnitude ranks them on the basis of L. Thus if a star A *appears* 100 times brighter to P than star B, then the apparent magnitude of B is 5 greater than the apparent magnitude of A. Of course, the way magnitudes are defined seems counterintuitive: the higher the absolute magnitude the weaker the star as a radiator and the weaker the apparent magnitude, the fainter the star appears to the observer.

The factor 100 chosen in the above example is especially simple since it corresponds to a difference of 5 magnitudes. How do we express an arbitrary factor as a magnitude difference? The answer is provided by the logarithmic scale which is ideally suited to measure differences expressed as ratinos.[1] Thus suppose star A has luminosity L_A and B has luminosity L_B. Then the absolute magnitudes of A and B, given by M_A and M_B, say, differ by

$$M_B - M_A = 2.5 \log \frac{L_A}{L_B}.$$

When $L_A/L_B = 100$, and its logarithm becomes 2. (10 raised to the power 2 gives 100). Therefore the right-hand side of the above equation

---

[1]See Appendix for a simple introduction to logarithms.

becomes equal to 5, as expected. Likewise we can write for apparent magnitudes m_A, m_B of A and B a similar relation based on their l-values:

$$m_B - m_A = 2.5 \log \frac{l_A}{l_B}.$$

Now we return to our question of how to ascribe an absolute magnitude to a star of given luminosity. First we settle the zero-point on the apparent magnitude scale. By convention we put $m_A = 0$ if $l_A = 2.48 \times 10^{-8}$ watts per square metre.[2] Then the above relation gives us the apparent magnitude m_B of any star whose l_B is known. A simple calculation of logarithms gives the aparent magnitude formula:

$$m_B = -2.5 \log l_B - 19.01.$$

where l_B is expressed in units of watts per square metre.

Now let us go back to the relationship between l_B and L_B, namely

$$l_B = \frac{L_B}{4 \pi d_B^2},$$

where d_B = distance of star B from us. To compare the luminosities of stars we must (theoretically) observe them all from the same distance. Again, by convention this distance is kept at 10 *parsecs*. A parsec is approximately 3×10^{16} metres. Its significance as a unit will become clearer in the next chapter when we discuss how star distances are measured. For the time being we just accept the parsec as a given unit of distance.

Then the rule is that the absolute magnitude M_B of star B is the apparent magnitude it would have *if its distance from us were 10 parsecs*. In other words, substitute $d_B = 10$ parsecs in the above formula, calculate the corresponding l_B and then convert it to m_B using the apparent magnitude formula. The answer you get is M_B. Naturally, to compute M_B for star B we need to know two things, its apparent magnitude m_B and its true distance d_B in parsecs from us. Given these quantities the above procedure gives:

$$M_B = m_B - 5 \log d_B + 5.$$

---

[2]This magnitude scale was proposed by Pogson in the 1850's. According to the above standard the stars Aldebaran and Altair are very nearly of first magnitude.

(As a check, if in the above formula we put $d_B = 10$ we then recover the result $m_B = M_B$.)

The Sun has absolute magnitude $M = 4.8$ approximately. Suppose it is moved to a distance of 2.7 parsecs, which happens to be the distance of star Sirius. The above formula tells us that the apparent magnitude of the Sun would then be $m = 1.96$. The apparent magnitude of Sirius itself is -1.42. This apparent magnitude difference of $1.96 - (-1.42) = 3.38$ implies that the ratio of apparent brightnesses of these stars is 21.5. That is, had the Sun been as far away as Sirius, it would have appeared 21.5 times fainter than Sirius. We thus discover that viewed from the Earth the Sun outshines all other stars simply because it is located very nearby. Had it been remote like other stars it would not have stood out in their company.

Colour index

At the end of Chapter 2 we hinted at the possibility that if a star's radiation is approximately like that of a black body then we can tell its surface temperature, because in a black body there is a connection between the radiation temperature and its overall colour. We now examine how this information can be actually utilized to determine the star's surface temperature.

To illustrate the method let us look at the intensity distributions of two stars. Figure 3.3(a) shows the distribution for star A and Figure 3.3(b) for star B. To recapitulate our discussion of Chapter 2, the intensity distribution tells us the amount of energy coming out of the star over all observed frequencies. This radiation, of course, is made of photons of different energies and Figure 2.9 showed such intensity distributions for black bodies of different temperatures. The intensity distributions can also be plotted with respect to wavelengths, by recalling from Chapter 2 that wavelengths of photons are related to their frequencies by the simple formula

$$\text{wavelength} \times \text{frequency} = \text{speed of light}.$$

Figures 3.3(a) and 3.3(b) are drawn with respect to wavelengths. Thus we can measure from them how much energy resides in radiation over a specified band of wavelengths.

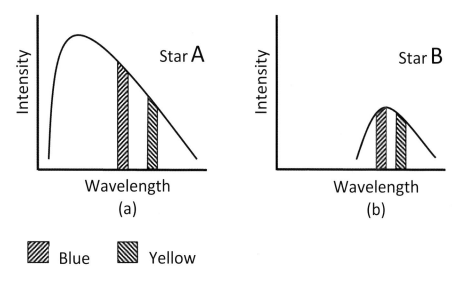

Figure 3.3: The relative intensities of blue and yellow lights are different for stars A and B.

In Figure 3.3(a) we see two wavelengths bands, one at the wavelengths of 5500–6000 Å which corresponds to yellow colour, and the other at the wavelengths of 4500–5000 Å which corresponds mainly to blue colour. Notice that the amount of energy in the two bands is not the same. Also, a similar exercise for star B in Figure 3.3(b) tells us the same thing, but with one difference. The ratio of energy in blue to the energy in yellow is higher for star A than for star B. Qualitatively we would notice this result through star A appearing bluer than star B.

Notice also that since the intensity curve of A peaks at a lower wavelength than does the curve for B, the surface temperature of A is expected to be higher than that of B. So qualitatively we arrive at the result that blue stars are hotter than yellow stars.

Quantitatively we can go further than this. Suppose we photograph the star image on a photographic plate using standard blue sensitive emulsion. If we try to determine the apparent magnitude of the star with this image by examining its brightness, we will discover that the main contributor to its illumination is the blue/violet radiation from the star. The magnitude so determined is called the *photographic magnitude* and is usually denoted by m_{pg}. We can also take the star's image on a green/yellow sensitive emulsion plate with a suitable filter that allows light waves of only this colour range to pass through. It is found that such an arrangement closely stimulates the response of the human eye which is most receptive to yellow colour. The magnitude of the star determined with such an image is called the *photovisual magnitude* and is denoted by m_{pv}. These two magnitudes effectively do what the two selected wavebands in Figure 3.3 did: they compare the relative intensity of the star's radiation in blue and yellow. Since the inverse square law of illumination works in the same way for all wavelengths, we do not have to worry about distance in this case. So, the difference

$$CI = m_{pg} - m_{pv}$$

will be larger for star B than for star A. The letters CI stand for *colour index*, an appropriate term since it tells us predominantly what colour is seen in the star's radiation.

For hot blue stars the colour index is large and negative while for cooler yellow-red stars it will be large and positive. The colour index therefore measures the star's surface temperature in a round-about way. In more modern terminology the symbols used for measuring apparent magnitudes in selected colour bands are U (for ultraviolet), B (for blue), V (for visual), J (for green) etc. The differences $B - V$, $U - B$, etc. thus measure the relative intensities of an astronomical object in these colour bands. If the star behaves very much like a black body of temperature T then it can be shown quantitatively that the following formula holds fairly well:

$$B - V = \frac{7300}{T} - 0.52.$$

Thus on the assumption of black body radiation the astronomer can tell the surface temperature of the star simply by measuring its colour index. In practice, however, he has to take account of any likely deviation from the black body spectrum before deciding upon the value of T. The surface temperature of the Sun so determined is close to 5800 K.

The Hertzsprung-Russell diagram

We are now in a position to appreciate the information contained in a typical H-R diagram. Basically the diagram plots the absolute magnitude of a star against its colour index. In other words, it tells us about the luminosity of a star and its surface temperature. But what it says about an individual star is not so significant as what it says about a group of stars.

In order to know the absolute magnitude of a star we must know its distance. Let us assume for the time being that we know it: in Chapter 4 we will in fact come to grips with that problem. The colour index, of course, does not depend on distance if we ignore interstellar extinction, which we shall do here.[3]

Figure 3.4 shows what the H-R diagram for a group of nearby stars typically looks like. The most obvious feature is that stars appear to congregate along a band AB, the top end A being in the region of luminous blue stars and the bottom end B being in the region of faint reddish stars. This band is called the *main sequence* and judging by its highly populated nature we presume that a typical star spends most of its life on the main sequence (see Figure 1.1 for an analogy concerning the period of adulthood in the human population). The Sun and a few well known stars are shown in the diagram.

Although the main sequence contains the vast majority of stars there are still an appreciable number in the upper right-hand region marked G and a few in the bottom left-hand region marked D. The former are *giant* stars which are very luminous but red in colour while the latter are faint *dwarf* stars. The adjectives "giant" and "dwarf" refer to the physical sizes of the stars and in due course we will discuss their detailed characteristics.

Since the colour index tells us the temperature at the surface of a star, Figure 3.4 also shows how this temperature decreases as we go towards the right in the H-R diagram. Thus dwarf stars are much hotter than giants. The giants, because of their predominantly reddish colour

---

[3]We have simplified the picture somewhat by ignoring the interstellar extinction caused by dust. This, as we will see in Chapter 4, does depend on wavelength. But we will ignore this effect here.

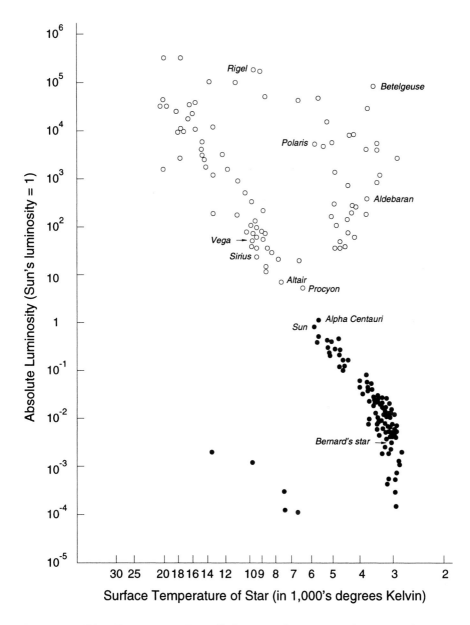

Figure 3.4: The Hertzsprung-Russell diagram showing nearby stars of various spectral classes. Colour indices have been converted to surface temperatures and absolute magnitudes to luminosities in relation to the Sun's luminosity.

Table 3.1: Spectral classes of stars.

Class Symbol	Distinguishing Features	Surface Temperature (K)
O	ionized helium	above 30,000
B	neutral helium	11,000–30,000
A	hydrogen	7200–11,000
F	ionized calcium	6000–7200
G	ionized calcium	5200–6000
	neutral metals	
K	neutral metals	3500–5200
M	neutral metals	below 3500
	absorption banks of molecules	
R	Cyanogen molecule band	below 3500
N	Carbon	below 3500

are also called *red giants*, while the dwarfs are commonly known as *white dwarfs*. Likewise, by looking at the vertical axis of the HR diagram we can argue that giants are bright and dwarfs are faint.

The H-R diagram incorporates another useful information inherent in the starlight, which we discussed in Chapter 2: the spectral features of stars which also vary from one star to another. Fortunately this variation can be related to the surface temperature of the star.

Table 3.1 shows how a sequence O, B, A, F, G, K, M, R, N can be formed with each letter characterizing stars of a particular class distinguished by their spectral lines. This sequence is in decreasing order of temperature with O stars the hottest and N stars the coolest.[4] In the hottest stars only lines of ionized helium and some other ionized elements are mainly seen. An atom becomes *ionized* when one or more of its orbiting electrons are stripped off, thus making the atom positively charged. The Sun belongs to class G which is characterized most conspicuously by the lines of ionized calcium.

---

[4]A scheme for remembering this sequence is based on a sentence constructed (except for the last two words) by Russell: *"Oh Be A Fine Girl, Kiss Me Right Now."* The last two classes R, N were included later.

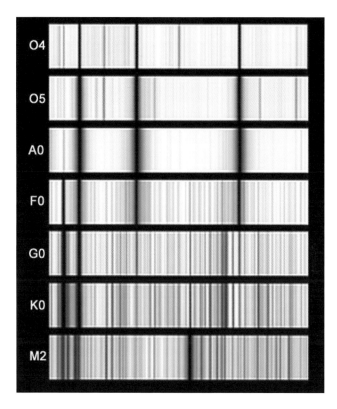

Figure 3.5: Star spectra showing prominent absorption lines corresponding to their spectral classes.

Figure 3.5 shows typical spectra of a number of stars belonging to different classes in the above sequence.

The classes are further subdivided into ten subclasses each and numbered A0, A1, . . . , etc. The Sun belongs to the subclass G2. This subclassification is based on finer details of a star's spectrum but we will avoid going into those details here.

It is important to remember that the spectral class does not uniquely place a star on the H-R digram. For example, we can have red giants also belonging to the same class G2 as the Sun. However, *if* we know that the star belongs to the main sequence then we can locate it there with the help of its spectral class.

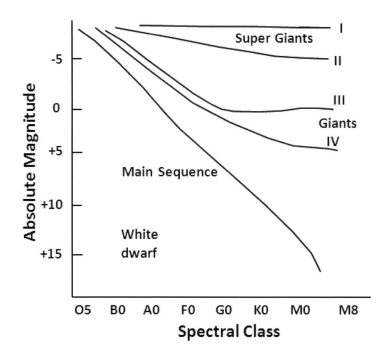

Figure 3.6: The H-R diagram with spectral and luminosity classes explicitly shown. Notice that we have additional type of stars called supergiants that are much more luminous than giants.

Just as spectral classes tell us a little more about what atomic processes are going on on the surface of a star, the *luminosity classes* tell us a little more about the star's luminosity. There are five luminosity classes I-V and these, along with the spectral classes are shown in Figure 3.6.

The stars of luminosity class I are the brightest and the largest in size, known as *supergiants*. Those of class II are the bright giants while stars of class III are the ordinary giants. Class IV includes those stars which are intermediate between the main sequence and the giant stars while class V are the main sequence stars.

The star can therefore be specified on the H-R diagram (provided it is a main sequence or a giant star) by its luminosity class and its spectral class. The former classification does not extend below the main sequence, to white dwarfs which are very faint anyway.

From facts to theory

The H-R diagram provides a framework in which we can place a star at any time of its life. So as the star evolves with growing age it may change its position in the H-R diagram. Our aim will be to formulate a theoretical framework to understand how this happens.

Except for one important issue, we have now completed the groundwork necessary to start wondering about such a paradigm. The important information still missing is embodied in the answer to the question "How far are the stars?" For, unless we know a star's distance from us, we cannot estimate its luminosity and until we know its luminosity we cannot place it at its rightful place on the H-R diagram. This is the question therefore that we shall address next.

Chapter 4

How Far are the Stars?

More than two thousand years ago, the most advanced astronomy in the world was practised in Greece. The Greeks had a simple view of the Universe. They assumed that the Earth was at the centre of the Universe, occupying a fixed and rigid position and the Sun and the stars were seen to move in circular paths from the East to the West because they were attached to the celestial sphere which revolved round the Earth (see Figure 4.1).

These Greek concepts in astronomy extended beyond Greece, to Arabia and even to India. The astronomical work of Ptolemy called *Syntaxis* was translated into Arabic and became known as *Al'magest* (the Arabs used the superlative "Al'majisti"), under which name it is now more commonly known even in Europe. The above geocentric view lasted in Europe until the revolution brought in by Nicolaus Copernicus in the sixteenth century. Copernicus argued that the Earth is not fixed. It spins about an axis and also orbits round the Sun.

However, India where the Greek tradition had taken root did have an independent thinker in Aryabhata. In his astronomical treatise *Arya-bhatiya* written in the 5th century A.D., Aryabhata put forward the alternative view that it is the Earth that spins about an axis and the stars are fixed in space. He gave the analogy of a boatman rowing down a river who finds the trees (which are fixed) on the bank going in the opposite direction. Likewise, argued Aryabhata, the stars rise in the East and set in the West because the Earth spins from west to east.

Figure 4.1: Photograph obtained by exposing the film for several hours to record the circular trajectories of the stars across the light. Today, we know that these apparent motions are due to the Earth's spin about its polar axis. (Photograph by David Malin.)

Figure 4.2: The montage of the Milky Way obtained by joining photographs of adjacent regions of the sky.

It is a reflection on how deep-rooted the Greek cosmology had become in India that Aryabhata, in spite of his prestige and influence, could not get this point across. Instead, his successors either ignored his thesis on the grounds that "he never said so" or interpreted it to mean something quite different!

Today the astronomer knows that Aryabhata had the right idea. Far from being "stuck" on a revolving celestial sphere, the stars that we see in the night sky are located at different distances from us and against their backdrop the Earth is seen to rotate about its polar axis.

Just how far are the stars?

The Milky Way

Before we discuss the methods used by astronomers to measure the distances of stars from us, let us look at the stellar picture as it is known and understood today.

If we look at the night sky we can make out a luminous belt across it wherein the density of stars is manifestly greater than outside it. This belt has long been known as the Milky Way (see Figure 4.2). Why are there more stars along this belt?

To understand the answer to this question imagine yourself (as in Figure 4.3) situated inside a not-too-thick disc full of stars. As you look out in the directions perpendicular to the disc you will encounter fewer stars than if you looked along the plane of the disc. This disc, seen from inside is projected on the celestial sphere (that is, the night sky) as a belt.

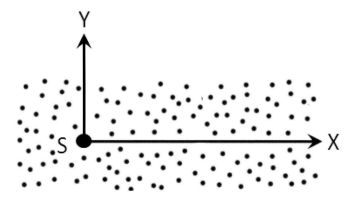

Figure 4.3: An observer S on the Solar System will see more stars in any direction SX lying in the plane of the galactic disc than in the perpendicular direction SY.

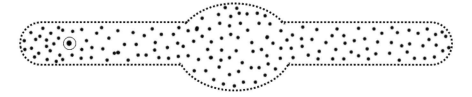

Figure 4.4: The Galaxy is like a disc with a central bulge. The Sun ($\odot$) is about two-thirds of the way from the centre. (Figure not drawn to scale.)

So the significance of the Milky Way is that we live inside a disc-type distribution of stars. A more careful survey of the stars in different directions gives two more points of information about this distribution:

1) There is a central bulge on the flat disc as shown in Figure 4.4.
2) We are *not* located at the centre of the disc.

The second point came as something of a blow to whatever ego man may have had during the post-Copernican era about his position in the Universe. Aryabhata, as we saw earlier, had already argued that it is not the celestial sphere that rotates round the Earth but, rather the reverse: the Earth rotates against a fixed celestial background. The work of Nicolaus Copernicus (1473–1543) and later astronomers like Johannes Kepler (1571–1630) showed that the Earth not only rotates

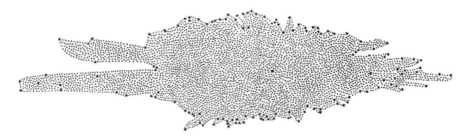

Figure 4.5: The Herschellian map of the Milky Way shows the Sun as a dark dot close to the centre of our Galaxy.

about its polar axis, but it also goes round the Sun in an elliptical orbit. Thus the central place in the Solar System belongs not to the Earth but to the Sun. When from their extensive observations of stars the father and son pair of astronomers William Herschel (1738–1822) and John Herschel (1792–1871) arrived at the disc-like structure of the Milky Way they thought that our solar system was at the centre of the disc. The Herschelian map of the Milky Way system is shown in Figure 4.5. This remained in force until the early twentieth century. However, by the early 1920's, detailed observational studies by Harlow Shapley (1885–1972) 'demoted' the Sun from its central position towards the periphery: as shown in Figure 4.5, the Sun is about two-thirds of the way out from the centre.

Figure 4.6 shows what the Milky Way system would look like when seen face on. Notice that it has at least two spiral arms winding outwards. The arms are regions where the star density is higher than elsewhere.

To get the Milky Way in proper perspective, the number of stars in the entire distribution is estimated to be two hundred billion. The disc itself has a diameter of about 100,000 light years and a thickness of about 10,000 light years. It is customary to refer to this vast system as "the Galaxy". Besides stars the Galaxy also contains gas and dust in diffuse form: but we will talk about these other components of the Galaxy later.

What is a light year? Light travels at a speed of about 300,000 kilometres per second. A *light second* is the distance covered by light in

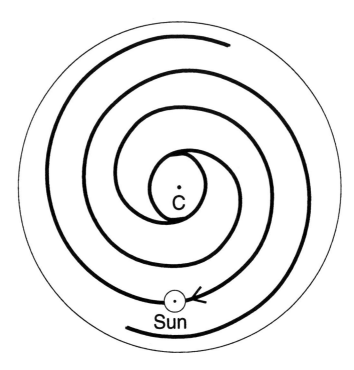

Figure 4.6: Two spiral arms are seen in the above schematic picture of our Galaxy.

one second, that is, a distance of 300,000 km. By a similar definition a *light minute* equals a distance of 18 million kilometres, a *light hour* equals a little over a billion kilometres while a *light year* is approximately 9.2 trillion km!

These figures convey some impression of the gigantic size of the Galaxy. If we were to undertake an imaginary journey from the Earth to some of the astronomical sites in the Galaxy, the times taken by such a tour are given in Table 4.1 below (provided, of course, that we can travel non-stop with the speed of light!).

Let us now proceed step by step in our search for the answer to the question "How far are the stars?" In short, we want to find out how the various distances mentioned in the above table were measured. We begin with the nearest star, the Sun.

Table 4.1: Approximate travel times within the Galaxy at the speed of light.

Tour	Time
Earth-Sun	8 minutes
Sun-Saturn	$1\frac{1}{4}$ hours
Sun-Pluto (edge of the Solar System)	$5\frac{1}{4}$ hours
Sun-Proxima Centauri (nearest star)	$4\frac{1}{4}$ years
Sun-Crab Nebula (see Figure 8.2)	6000 years
Sun-Galactic Centre	30,000 years
Galactic Centre — Edge of the Galaxy	50,000 years
complete round of the Galactic perimeter	300,000 years

Distance of the Sun

A method which is extensively used in the measurement of astronomical distances is illustrated by an example in Figure 4.7. Here we have a lighthouse on an island, visible from the mainland. We want to find out how far it is from us, *without having to leave the mainland*, since this is precisely analogous to our astronomical problem.

As shown in Figure 4.7 the problem is solved by the method of *triangulation*. Observe the top C of the lighthouse CD from two vantage points A and B on the mainland. The line AB is chosen to pass through D the base of the light house. The distance AB can be measured as well as the angles CAB and CBA. It is then a straightforward matter to complete the triangle ABC on paper with suitable scale for AB. We then know where C is located in relation to A and B and hence its distance from any point on the mainland. The line AB is called the *baseline* for this triangulation.

Since the astronomical objects are far away, we need to have a long baseline. For, if we don't, our triangle ABC will be very "thin", in the sense that the angle ACB will be very small. A small error in the measurement of this angle results in a large error in the estimate of the distance to the lighthouse.

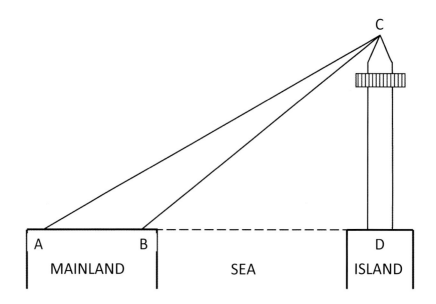

Figure 4.7: How far is the lighthouse CD from mainland?

Of course, the longest baseline we can get on the Earth is by observing from the ends of its diameter. These points are separated by a distance of about 12,800 km. Large though this distance is, it is not large enough to give the Sun's distance from us accurately. But it can be used to determine the distance of Mars when it is at its closest separation from the Earth. This happens when the Sun (S), the Earth (E) and Mars (M) are in a straight line with the Earth in between. Figure 4.8 illustrates this special situation. The distance EM can then be measured by triangulation.

But how do we measure SE, the Sun-Earth distance? For this we use Kepler's laws of planetary motion. Kepler's third law states that the *square* of the time period taken by the planet to go round the Sun varies in proportion to the *cube* of its distance from the Sun. The Martian period is 687 days, while the Earth's period is $365\frac{1}{4}$ days. From Kepler's law we therefore get the relation

$$\left[\frac{SM}{SE}\right]^3 = \left[\frac{687}{365\frac{1}{4}}\right]^2.$$

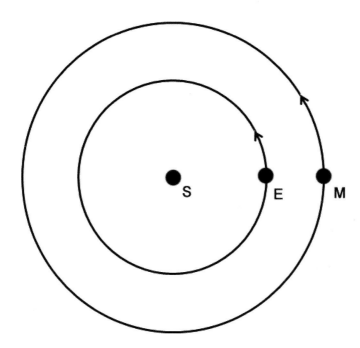

Figure 4.8: The Earth, the Sun and Mars are shown here to be on a straight line. This situation is used for measuring the Sun-Earth distance, as described in the text.

Solving this equation we get the result SM = (3/2) SE, approximately. Thus SE = 2EM, and we thus determine SE from our measurement of EM.

Of course we have illustrated the method of measurement by simplifying it somewhat. In reality the planets move in elliptical orbits round the Sun and the mathematical problem of determining SE is more complicated than for the circular orbits of Figure 4.8. However, the principle behind the measurement is conveyed correctly by our example.

Modern technology has provided a better technique than this older triangulation method using Mars. By bouncing radar signals off the planet Venus (V) when it comes between the Earth (E) and the Sun, we can measure its distance from the Earth directly. For, the radar is a form of microwaves which (as we saw in Chapter 2) travel with the speed of light. Since the total time for the signal to travel to Venus and echo back to the Earth is 300 seconds, we conclude that the one-way distance EV is *half* the range covered by light in this period.

The first such radar experiment was performed in 1958 at the Lincoln Laboratories of the Massachusetts Institute of Technology. It has since been performed by many laboratories with better and better accuracy. The Sun's distance from the Earth is now known to be 149597870.61 km, with an uncertainty of about 100 metres. This distance is called the *Astronomical Unit* (AU).

Stellar distances by triangulation

Having determined the Sun-Earth distance we are now in a position to use the Earth's orbit as a baseline for triangulation of stars. Figure 4.9 illustrates the method, again simplified here to the assumption of a circular orbit.

E_1 and E_2 are two positions of the Earth on its orbit at an interval of six months. Thus $E_1 E_2$ is the diameter of the orbit with the length of approximately 300 million km. Σ is the star whose distance we want to measure. Provided the star is not very far away, we can measure the angles $\Sigma E_1 E_2$ and $\Sigma E_2 E_1$ with reasonable accuracy. As discussed earlier, the triangle is a narrow one and hence the measurements of distances ΣE_1 and ΣE_2 can be in considerable error if the angles are not accurately measured.

During the Earth's motion along its orbit round the Sun, the star's direction will therefore appear to change in a continuous manner. As shown in Figure 4.9, the apparent position of the star as observed from the Earth will be seen to describe a small ellipse. The angle $E_1 \Sigma E_2$ will be largest when both sides of the triangle, $E_1 \Sigma$ and $E_2 \Sigma$ are equal. Suppose at this stage this angle is $2p$. Then p is the angle subtended at the star by the Earth-Sun line held perpendicular to the line of sight.

This angle p is called the *parallax* of the star. The distance of the star from the Earth-Sun line at which p is as small as 1 arc second ($= 3600$th part of a degree) is called a *parsec*. One parsec equals approximately $3\frac{1}{4}$ light years; in terms of terrestrial units 1 parsec equals 30 trillion kilometres approximately. This was the distance unit introduced

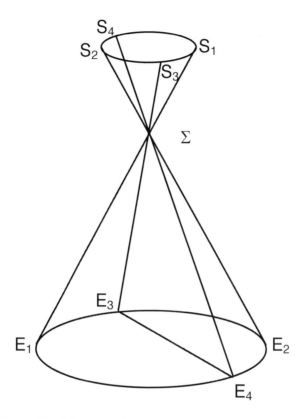

Figure 4.9: A simplified diagram to illustrate the principle of parallax of star Σ. The Earth's orbit is assumed circular and $E_1 E_2$ is its diameter with property that $E_1 \Sigma = E_2 \Sigma$. $E_3 E_4$ is any other diameter of the Earth's orbit. S_1, S_2, S_3, S_4 are the projections of the star on the celestial sphere when viewed from E_1, E_2, E_3, E_4 respectively. These projected points lie on an ellipse with major axis given by $S_1 S_2$. The parallax angle is half the angle $E_1 \Sigma E_2$, that is half the maximum change in the star's apparent direction during the entire orbit.

in Chapter 3 and we now see how it arises naturally in stellar astronomy. Provided p is not very small, say not less than 0.05 arc second, the star's distance can be estimated by the triangulation method reasonably accurately (say with less than 10 percent error). Some 700 stars satisfy this criterion, although parallaxes have been measured for thousands of stars.

There is, however an apparent paradox here! The 700 or so stars mentioned above lie within distances of 20 parsecs. But most of them are not very powerful radiators of light and so are invisible to the naked eye. They are observed with photographic aids and a telescope. On the other hand the stars seen in the sky with the naked eye are generally located much farther than 20 parsecs so that their distances cannot be measured accurately by the parallax method. These stars are visible because they are intrinsically very luminous whereas the 700 stars mentioned above are near but intrinsically very weak. Thus we should not be misled into believing that bright-looking stars are necessarily near us and faint-looking stars are necessarily far away. A star's intrinsic power also plays an important role in the overall problem of estimating its distance.

As a throwback to history, we should mention the Greek astronomer Aristarchus (310–230 BC approximately) who had the correct idea that the Earth travels round the Sun. This was against the commonly believed 'fixed earth' hypothesis and Aristarchus was asked to provide proof for his idea. He provided the following argument. Since the planet changes its position as it orbits the Sun, Aristarchus correctly argued that the direction of a foreground star would appear to change with respect to the background of more distant stars and this effect should be measurable. Today we see that he was referring to the above parallax-based measurement. Aristarchus, however, expected the stellar distances to be much less than a parsec or a light year and so he estimated somewhat large parallax angles. The measurement techniques in those days were rather primitive with the result that the expected parallaxes were not found. This negative result led to reinforcing the prevailing belief that the Earth was fixed in space.

The use of the H-R diagram

The inverse square law of illumination together with the H-R diagram can be used to measure distances of stars which are too far for the triangulation method to be of any use.

Suppose, for example, that we know the spectral class of a main sequence star from its spectrum, to be G2, the same as that of the Sun (see Chapter 3). We know what the absolute luminosity of the Sun is. From the uniqueness of the Sun's position on the H-R diagram, we would argue that a star with the same spectral class as the Sun should occupy the same position as the Sun on the H-R diagram. We therefore know what the absolute luminosity and hence the absolute magnitude (M) of the star is, while from observations we can directly measure its apparent magnitude (m). Then with the help of the formula obtained in Chapter 3 the distance d of the star is given by solving the equation

$$5 \log d = m - M + 5.$$

Note that, as defined in Chapter 3, this distance is measured in parsecs.

There is, however, a snag in this procedure. It assumes that the star is, like the Sun, situated on the main sequence. In Chapter 3 we saw that a G2 star could also be a red giant, in which case, compared to the Sun, its absolute luminosity would be about 100 times higher and its absolute magnitude M lower by 5. Therefore it is necessary to use this test with caution. Preferably it may be used for a group of stars.

The H-R diagram can be used to measure the distance of a remote star cluster as per this hypothetical example.

Figure 4.10 shows the plot of the apparent magnitude against the colour index of the stars in a cluster. The plot looks like the main sequence of an H-R diagram but is not quite the same. The difference is that the H-R diagram plots absolute magnitudes whereas Figure 4.10 plots the apparent magnitudes. We can convert one to the other provided we know the distance of each star in the cluster. Although the distances vary from star to star, this variation is not much in a remote cluster. This is like saying that the distance of a tree is 100 metres from us when we know that the individual distances of all its leaves are not exactly 100 metres; some may be more, some may be less: but this variation hardly matters.

So let us assume that the *average* cluster distance from us is d parsecs. Then we can convert the apparent magnitudes of Figure 4.10 into absolute magnitudes by subtracting $5 \log d - 5$ from them. How much is $5 \log d$? Again taking the example of the Sun, we know that it has absolute magnitude $\simeq 5$ and colour index 0.6, on the H-R diagram. In Figure 4.10, the colour index 0.6 seems to correspond to an average apparent magnitude of 10. The difference $m - M$ is therefore $10 - 5 = 5$.

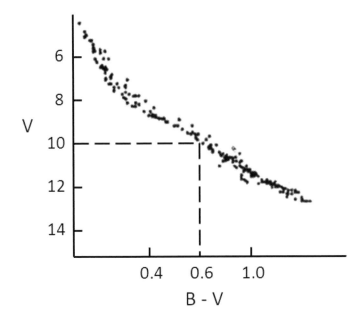

Figure 4.10: The plot of apparent magnitude against colour index for a star group.

We therefore get

$$5 \log d - 5 = 5,$$

that is, $\log d = 2$. This means that the cluster is at a distance of 100 parsecs. Alternatively, imagine Figure 4.10 plotted on a transparent sheet which is then placed on an H-R diagram of stars of known distances. The sheet is made to slide up or down until the two main sequences coincide. The difference $m - M$ can then be read off directly from the magnitude scales of the two diagrams.

Powerful though this method is for measuring stellar distances it can give quite misleading results if we ignore an important factor. This factor is known as interstellar extinction and its effect can be qualitatively seen with the following example.

We have already noted that the inverse square law of illumination implies that the more remote a source light is, the fainter it appears. As we recede from a street light we find it getting fainter and fainter. There is, however, another cause that can make even a comparatively nearby street light appear faint. If there is thick fog around, the light will be dimmed even though it is not very far away because it is absorbed and scattered by fog particles.

Likewise, scattering and absorption in the Galaxy will tend to dim starlight coming towards us. A star will therefore appear dimmer to us than it would have been if no fogging material were present en route. The apparent magnitude of such a star will therefore be higher than it would have been in the absence of en route extinction. If we used our formula above as it stands, we would therefore tend to *overestimate* the star's distance.

As late as the beginning of the twentieth century the phenomenon of interstellar extinction was not known. If we look back at the montage of the Milky Way in Figure 4.2 we notice dark patches interspersed with bright patches. If the bright patches are due to starlight what are the dark patches due to? Earlier it was thought that these patches represent *absence* of stars. Now it is realized that stars are present even in those dark regions but their light is unable to reach us because of intervening dust. This interstellar extinction of light can work in varying degrees, from generally dimming the star's image to totally extinguishing it. The Horsehead Nebula shown in Figure 4.11 is a classic example of a cloud of interstellar dust.

Thus many of the earlier estimates of stellar distances in our Galaxy had to be revised *downwards* to take into account the effect of interstellar pollution! This was also the reason why the early investigators like the Herschels felt that the Galaxy as a whole was centred round the Sun. For, with their more limited observing aids they failed to see stars beyond a certain distance which is roughly the same in all directions. They could not, for example, see the real centre of the Galaxy because light from so far away was almost totally extinguished by the time it reached Earth.

Of course the extinction is not as strong at longer wavelengths as it is for visible light, and this is why we are able to ascertain the overall picture of the Galaxy. In particular the surveys of the Galaxy using the waves of 21 cm wavelength have played a very important role in our understanding of its structure (see Figure 4.12).

Figure 4.11: The Horsehead Nebula.

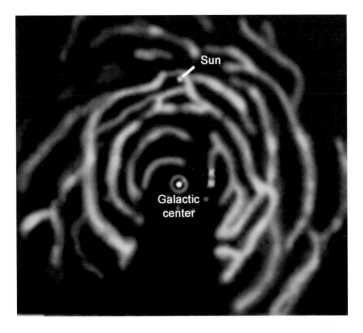

Figure 4.12: A map of the Galaxy with 21-cm waves. The spiral arms are apparent through the denser regions.

It should be noted that the gas in the Galaxy does not play a major part in the light extinction process. The culprits are believed to be solid particles of matter which are far more efficient than the atoms or molecules in the gas in extinguishing starlight either by absorbing it or by scattering it in different directions. Theoreticians can calculate how much extinction a solid grain of a particular size and of a particular material can produce at different wavelengths and these investigations have led to models of what the interstellar dust is made of. Solid hydrogen, graphite, silicates are some of the popular candidates and the typical grain size is measured in microns (1 micron = one thousandth part of a millimetre). The test of a model lies in its ability to correctly reproduce the observed extinction at different wavelengths of a star's light.

Once astronomers knew the degree of extinction they could correct the earlier overestimates of astronomical distances of remote stars. Thus it was that by the 1920s more accurate dimensions of the Galaxy became available to the astronomers.

There are several other methods for measuring stellar distances which make use of some special property or other of the type of star or star clusters under study. One method which uses variable stars (that is, stars whose luminosity varies up and down with time) will be described later. We will not go into other methods here, since our main purpose in this chapter was to outline the *general* methods that are used for measuring the distances of stars in the Galaxy. These methods have produced reasonably accurate distances of nearby stars, say up to 20 parsecs and approximate distances of more remote stars. The accuracy of any method inevitably diminishes as we look at more and more remote stars.

Inputs from Space Technology

Ground-based measurements suffer from limitations like dimming by the atmosphere and imperfect seeing because of air currents in the atmosphere making the image unsteady. By going above the atmosphere, one can improve on both these counts besides having the advantage of twenty four hour observing. Hence stellar astronomers have used

Figure 4.13: Hipparcos Satellite.

satellite based telescopes for improving and extending their data on parallax measurements of stars. Also, if one has accurate spectroscopy, one can also measure motions of stars.

The first such experiment was the Hipparcos satellite "**Hi**gh **pr**e-cision **par**allax **co**llecting **s**atellite" with the acronym commemorating Hipparchus, the distinguished Greek mathematician and astronomer of two millennia ago (160–125 BC). The satellite (see Figure 4.13) was launched by the European Space Agency in 1989 and it was collecting data and processing it until 1993. The data was published as a catalogue of some 100,000 stars in the Galaxy, giving their parallaxes and also their proper motions. The proper motion is the motion of the heavenly object perpendicular to the radial line of sight from the observer to the source. Later the Tycho 2 Catalogue of 2.5 million stars was published in the year 2000. This effort has added considerably to the observational base of the stellar astronomer.

Now that we know how far the stars are and how luminous they are, we can embark on the main purpose of this book: to study how the structure of a star changes with its age, and how these changes are detected through astronomical observations and through the H-R diagram.

Chapter 5

A Star is Born

We now take up the story of a star's life, beginning with its birth. The story is not yet complete as there are still several questions left to be answered. But a broad paradigm has emerged from the various observations and discussions amongst theoreticians. According to the current theories of star formation, our story begins in a dark interstellar cloud of gas and dust.

Figure 5.1 has the image of the Orion Nebula. It is a cloud of gas in which star formation is believed to be taking place. The bright part of the cloud is lit up by the light from newly born stars. There is, however, more to this cloud than what meets the eye. Considerable information is hidden from us if we just rely on the light from the cloud coming through the visual window. As the techniques of microwave and infrared astronomy improved in the 1960's and 1970's the interstellar clouds began to reveal much of their hidden contents.

Giant molecular clouds

The Orion Nebula and its dark surroundings shown in Figure 5.1 form what is known as a Giant Molecular Cloud (GMC). "Giant", because it has a very large size, of the order of 100 light years! "Molecular" because it is found to contain molecules — combinations of atoms in the form of chemical compounds. Each molecule like the atoms we discussed in Chapter 2, has its characteristic energy levels. The energy levels of

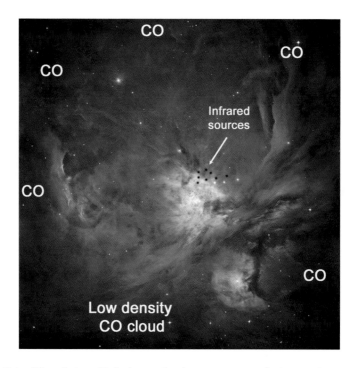

Figure 5.1: The Orion Nebula with the presence of the carbon monoxide molecule indicated by letters CO. Star formation is taking place in the region of infra-red sources.

interest to the astronomers are those arising from internal rotations of the molecule. And, as in the case of atoms, when a molecule changes its rotational state it does so either by absorbing radiation and going to a state of higher energy or by emitting radiation and descending to a state of lower energy.

The characteristic frequencies of molecular radiation are considerably smaller than those of radiation from atoms; they are of the order of 10^{11} cycles per second. This radiation as we see from Table 2.2, lies in the microwave region. By tuning the antenna to the characteristic frequency of a given molecule, the astronomer can detect the presence and the density of that molecule in the cloud. Figure 5.2 shows a microwave receiver of 36-feet diameter used for such observations. It operates at wavelengths longer than 1 millimetre.

Figure 5.2: A millimetre-wave telescope at Kitt Peak, Arizona for detection of molecular lines in interstellar clouds.

In the mid-1950s the visionary astronomer Fred Hoyle thought of such clouds as existing in the interstellar space, clouds containing molecules. He thought of this idea in the broader concept of star formation. In those times radio astronomy was emerging as a new and powerful branch of observational astronomy and radio techniques had revealed the presence of neutral atomic hydrogen all over our Galaxy.

It is known that the neutral hydrogen atom can exist in two states. The state of higher energy has the electron and proton in the atom spinning in parallel directions, while in the second state their spins are in *anti*parallel directions. Figure 5.3 illustrates the situation. As per nature's laws, the atom in the higher energy state can *spontaneously jump down* to the state of lower energy. The energy difference is radiated by the atom as a quantum of radiation. Theoretical calculation shows that this quantum corresponds to a wavelength of radiation approximately equal to 21 centimetres. This wavelength falls in the radio range

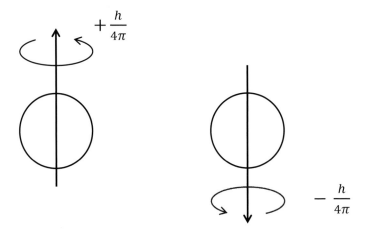

Figure 5.3: In (i) we have the electron and proton of the hydrogen atom spinning in parallel directions while in (ii) they are spinning in opposite directions. The second state is of lower energy.

as per Table 2.2. That is how astronomers could claim to have proved the presence of neutral hydrogen anywhere this radiation came from.

However, when Hoyle proposed that larger chemical structures, molecules made of several atoms exist in large cloudlike structures in the Galaxy, he discovered that it was almost impossible to get his fellow-astronomers to accept this idea. Thus he had his papers rejected by reputed research journals. He then thought of another medium for popularising this idea: *He wrote a science fiction novel centred around such a molecular cloud.* Known as *The Black Cloud* the novel became very popular. And, as mentioned earlier, the notion of molecular clouds received confirmation once antennas receiving and detecting millimetre wavelength radiation began to come on the line from the 1960s.

Table 5.1 below gives some idea of the wealth of data received this way. The interstellar space is by no means uninteresting if the contents of the GMC's are anything to go by. Notice that not only inorganic molecules are present but also organic ones. The fact that many of the latter are found as components of the basic biological molecule, the DNA, raises the intriguing question: Can life systems exist in space given that the basic building blocks are there?

Table 5.1: Molecules in space*.

Number of Atoms in the Molecule	Inorganic Molecules	Organic Molecules
2	H_2 (hydrogen)	CH (methylidyne)
	OH (hydroxyl)	CN (cyanogen)
	SiO (silicon monoxide)	CO (carbon monoxide)
	NS (nitrogen sulphide)	CS (carbon mono-sulphide)
3	J_2O (water)	HCN (hydrogen cyanide)
	H_2S (hydrogen sulphide)	HCO (formyl)
	SO_2 (sulphur dioxide)	HNO (nitroxyl)
4	NH_3 (ammonia)	H_2CO (formaldehyde)
	—	HNCO (isocyanic acid)
5	—	H_2CHN (methanimine)
	—	HCOOH (formic acid)
6	—	CH_3OH (methanol)
	—	$HCONH_2$ (formamide)
7	—	CH_3NH_2 (methylamine)
8	—	$HCOOCH_3$ (methyl formate)
9	—	$(CH_3)_2O$ (dimethyl ether)

* This is a partial list to give a flavour only!

Of course, each molecule is distributed in its own different fashion. The CO molecule (Carbon Monoxide) for example gives the largest scale distribution of the GMC. The CO-picture of the GMC containing the Orion Nebula of Figure 5.1 extends well beyond the optical picture. CO has been found in other parts of the Galaxy and in other galaxies as well.

A GMC is by no means homogeneous in composition. As indicated in Figure 5.4 it has inhomogeneities on differing scales. Always we find a very dense pocket of matter surrounded by a less dense shell which in turn is surrounded by an even less dense shell and so on. The small dense region shown in Figure 5.1 is called a *molecular cloud* (without the adjective "giant") and has a diameter of about a light year. It is these compact dense regions that hold the clue to star formation.

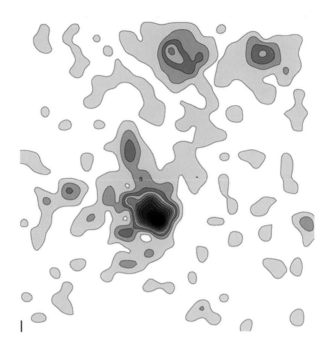

Figure 5.4: A GMC develops inhomogeneities shown here by contours, with the innermost region (filled contour) being the densest. This is where protostars develop.

The formation of protostars

What is a star? A star is after all a ball of hot dense gas which is capable of shining through radiation of energy. To make a star therefore, we need to compress a region of molecular cloud sufficiently hard until it becomes dense and hot enough for this purpose. This compression is achieved by the force of gravity. As we shall discover again and again in the course of this tale of stars, gravity plays a crucial role almost all through a star's life.

So far as the GMC is concerned, any initial inhomogeneity in it tends to increase as a result of gravity; for denser regions have stronger gravitational pull on the surroundings and therefore tend to pull in more matter and become denser still. This is how the inhomogeneities of Figure 5.4 develop.

The role of gravity in the contraction with the GMC may be likened to the discovery of some precious natural resource like oil in an undeveloped country. The discovery attracts influx of people from the surroundings leading to greater economic activity in the region. Thus a disparity develops between that region and its surroundings, a disparity that keeps on growing. However, this cannot go on forever since restoring socio-economic forces begin to assert themselves and the region finally settles down to a steady state. In the contracting cloud also, opposing forces begin to appear and eventually lead to a steady situation. This comes about in the following way.

When a gas is compressed it heats up and if it is hot enough it begins to radiate heat and light. This radiation as well as the increase in the chaotic motion of gas molecules and atoms (see Figure 5.5) generate

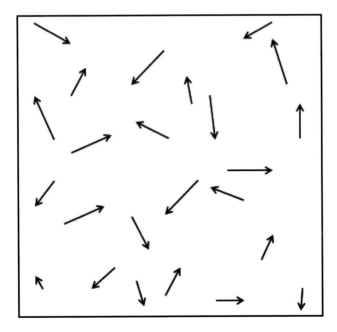

Figure 5.5: The arrows indicate directions of motion of gas particles. The magnitudes as well as directions of particle velocities are uncoordinated, i.e., chaotic. The extent of chaotic activity is reflected in the overall temperature of the gas.

pressures which oppose the gravity induced contracting tendency of the molecular cloud. The temperature and pressure are the highest in the centre and the lowest on the periphery.

One of the basic laws of heat is that it flows from a zone of high temperature to a zone of low temperature, provided avenues are available for the heat to flow. In the *protostar*, that is, in the molecular cloud described so far, which is on its way to become a star, there are two ways in which heat can flow from the hot central zone to the cooler peripheral regions. One way, called *convection* has the hot gas particles from the centre bodily moving outwards to the cooler regions, much in the same fashion that water rises upwards when heated to the boiling point. In the other method of heat transport the carriers are photons, the particles of light (see Chapter 2). The photons also travel outwards bearing energy which appears as heat and this process is naturally known as *radiation*.

Not both processes are always equally efficient: at any stage which of the two dominates in a particular part of the star may depend on several physical circumstances. Convection for example can be inhibited if gas particles find it more and more difficult to travel outwards as can happen if the central densities in the star become very high. Likewise, radiation becomes inefficient if the light photons are too frequently scattered by the protostellar material and are thus prevented from taking the straight and narrow path outwards. We will encounter this competition between the two modes of heat transport later also, when (in the following chapter) we discuss the internal structure of a properly formed star.

To return to our protostar, we find that in the early stages the convective mode operates well and efficiently (unless the protostar is too massive, say more than three times as massive as the Sun). The result is that heat is rapidly transported outside and radiated away from the surface of the cloud. The cloud is therefore very luminous to begin with.

This luminous phase does not last for long, however. For, to supply the energy needed for radiation the protostar has at this stage only one source available: its gravitational energy reservoir. To draw on this energy the star must contract rapidly. We will discuss in the next chapter in somewhat greater detail how gravitational contraction releases gravitational energy. For the time being let us take the idea for granted and explore its consequences for the molecular cloud. The protostar contracts rapidly in the early stages and is able to maintain the large flow of energy from the centre to the periphery and from thence into outer space. But as it contracts and becomes dense, convection ceases to be an efficient transporter of energy and so the protostar becomes less and less luminous. It also tends to contract more slowly now compared to its early phase. The contraction and convective heat transport therefore go on more and more slowly till a certain stage when convection ceases to be important.

This phase in the protostar's life is called the *Hayashi phase* because it was first discussed in detail in 1966 by the Japanese astronomer, Chushiro Hayashi. The phase has one important bearing on the protostar's outward appearance which we have so far not discussed. The remarkable result is that during this contraction the surface of the protostar maintains a constant temperature of around 4000 K.

The reason is as follows. At this temperature the internal motions in the gas are so rapid that the atomic electrons get knocked out of their orbits around their respective atomic nuclei. The electrostatic attraction of the nucleus is no longer able to hold the electrons which break loose once the temperature goes beyond 4000 K. These free electrons are remarkably efficient in scattering any outgoing radiation in the protostar. Below 4000 K, the electrons are bound within the atoms and cannot do much to the radiation which therefore streams through a swarm of intervening atoms. Figure 5.6 shows how this temperature effectively fixes the protostar's surface: within it the temperature is higher than 4000 K and the radiation is trapped, outside it the temperature is lower than 4000 K and the radiation streams out. The observer therefore sees a ball of radiation with a surface temperature of 4000 K.

Figure 5.7 illustrates the Hayashi phase on the H-R diagram. It is a straight track shown by the line *SH* (often called the *Hayashi track*) at the constant temperature of 4000 K. The point *S* marks the highly luminous phase in the beginning and the point *H*, the end of the Hayashi phase when convection has more or less ceased to be important. Notice also the times shown at different stages on the line *SH*. These times indicate that the contraction was rapid to begin with but slowed down later.

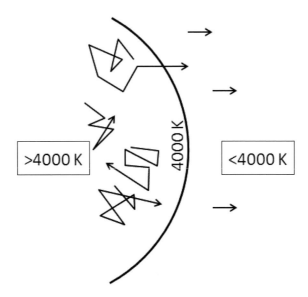

Figure 5.6: Photons in the inner, hotter ($\sim$ 4000 K) region are trapped while those which leak past the 4000 K surface into cooler outer region escape. The critical 4000 K surface is thus the effective surface of the protostar.

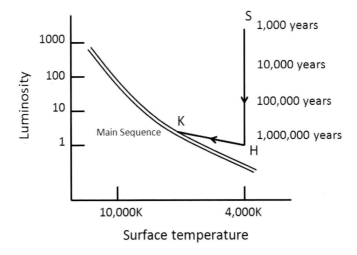

Figure 5.7: The Hayashi track SH shown on the H-R diagram with luminosity (on arbitrary scale) plotted against the surface temperature.

The surface temperature of 4000 K means that the protostar is hardly visible in the optical range. It is, however, very bright in the infrared at least in the early stages. The studies of the Orion Nebula have indeed revealed copious infrared emission from the part shown by the arrow in Figure 5.1. This is the main reason for believing that new stars are being born there. Since the time spent by the protostar on the Hayashi track is only a few million years (even less if we restrict it to the initially luminous phase) which is a small fraction of the star's entire life, we may liken the Hayashi phase to the toddler stage in human life.

The "toddler" protostar at the end of the Hayashi phase has next to rely on the radiative mode of heat transport from the centre to the periphery. The protostar of course continues to contract and it gets hotter and hotter at the centre. Its luminosity increases slightly since radiative transport is able to improve somewhat on the convective mode at its worst. On the H-R diagram therefore the protostar moves along the track HK. The surface temperature rises above 4000 K, but it has a more spectacular increase in the centre. It is this latter phenomenon that eventually marks, at K, the beginning of the star's life as an "adult" member of the population when the star begins generating its own energy. *The point K brings the star to the main sequence.*

We will postpone our discussion of this entirely new situation to the following chapter. Here we turn our attention to another important event that may happen at the time a star is born.

The formation of planets

Although we have described the early stages in the star's life as if it were an isolated cloud, what we have to remember is that the cloud is a subunit of the GMC, which according to Figure 5.4 is full of inhomogeneities as if it were fragmented into lumps. Each lump is a protostar; thus stars form not in isolation but in groups.

The GMC as a whole and certainly its subunits which condense into stars may be rotating, just as the Earth is found to rotate about an axis. However, contrary to what happens to the Earth which is a rigid body, a gas cloud finds that as it shrinks in size, its rate of spinning about its axis increases. This is the same effect under which an ice-skater finds

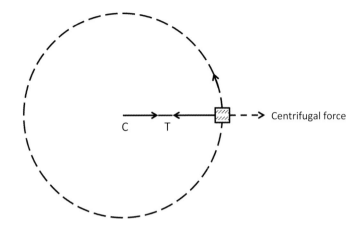

Figure 5.8: The stone whirled round in a circle is always pulled to the centre C by the tension T in the string. In the frame of reference of the stone, this tension balances the outward centrifugal force to keep the stone at fixed distance from C.

herself spinning more rapidly if she brings her hands inwards. Known as the conservation of angular momentum, under this effect the gas cloud spins more rapidly as it shrinks. Also, because of rotation it is not able to condense as a spherical ball. For, rotation generates a kind of fictitious force called the *centrifugal force* that plays an important role in the condensation process.

Figure 5.8 illustrates the nature of this force. It shows a stone tied to a piece of string and whirled around in a circle. The person who does the whirling job feels a tension in the string. It arises because the string is trying to pull the stone towards the centre of the circle while the stone has a tendency to fly away from the centre. If the string breaks, the stone will indeed fly away. The reason for tendency on the part of the stone is explained by Newton's laws of motion. If the string were cut the stone is free and as a free particle it will move in a straight line with a fixed velocity. This is Newton's first law of motion. Seen from the centre of the circle it tends to fly away in a tangential direction to its circular path. The centrifugal force is a concept that describes this tendency. It is *not* a real force. To an observer sitting on the stone, the tension of the string is towards the centre of the circle; and to explain

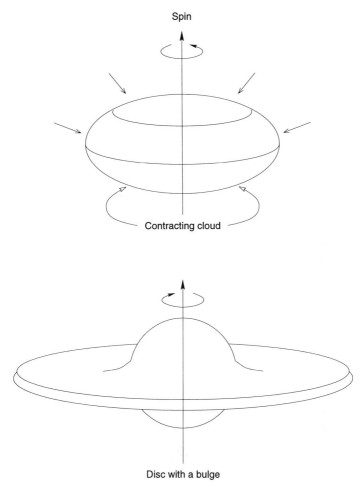

Figure 5.9: The above stages show how an initially spherical rotating cloud bulges out with a disc.

his apparent state of rest, the observer needs to postulate this force in the opposite, i.e., outward direction from the centre of the circle.

In Figure 5.9 we find that because of this "fly away" tendency, as the gas cloud contracts, it spreads out in the directions away from the axis of rotation. The result is that it takes the shape of a disc surrounding a central bulge.

It was the French physicist and mathematician, Laplace early in the 19th century who argued that the Sun and the planets may have been formed from such a contracting and rotating cloud, with the central bulge forming the Sun and the disc forming the planets.

The picture needs some amplification, however. It is found, for example, that the Sun does rotate about an axis perpendicular to the plane in which the planets move; but it does not rotate as rapidly as the above picture would lead us to believe. Moreover, the disc in the above picture does not extend far enough away to account for all the planets. Some new ingredient is needed that would not only slow the rotation of the central bulge but would also extend the disc much farther away.

It was pointed out independently by H. Alfven and F. Hoyle in the 1940's that the extra ingredient in the above recipe is the magnetic field. Figure 5.10 describes how the magnetic field might do the job.

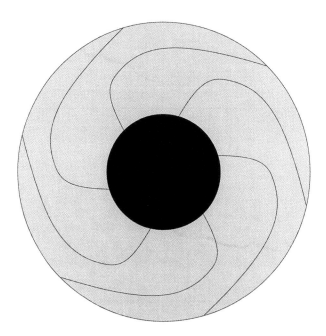

Figure 5.10: Winding magnetic lines of force joining the points on the central bulge to those on the disc above tend to slow down the inner region while making the outer disc rotate faster.

Consider two points of the rotating cloud: A, a point in the central bulge, and B, a point in the disc. The magnetic field would link A and B by a *line of force*. The line of force is an imaginary line in the space which denotes the direction of the magnetic field at any point on it. It is easy to trace the lines of force of a bar magnet. Place the magnet on a cardboard and sprinkle iron filings around it. By tapping the cardboard gently we find that the filings settle down to the pattern shown in Figure 5.11. The lines of the pattern are the lines of force. A typical line of force shows the direction of the magnetic field at any point on it as along the tangent at that point.

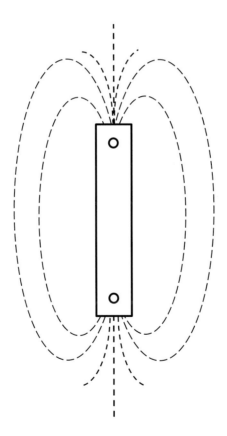

Figure 5.11: The iron filings lie along the lines of force for the bar magnet shown above. Only a few typical lines are shown.

But to return to Figure 5.10, imagine two typical cloud particles at
A and B. The lines of force have a tendency to stick on to the particles
in the cloud all the time. So as A and B rotate about a common axis,
the line of force moves with them. But, since A and B do not rotate
at the same rate with A rotating faster than B, the line of force gets
twisted. And when twisted it tends to straighten itself.

In this process it pulls A back while making B go faster, thus revers-
ing the tendency of the cloud as a whole to rotate faster in the centre
and slower in the outskirts. This resistance offered by the line of force
serves the two required needs: 1) it slows the central part down, 2) it
pushes the outer parts further out by making them rotate faster.

It is believed that the fast rotating thin disc cannot maintain its
shape for long. It breaks up into large and small lumps of matter which
eventually become the various components of the planetary system:
planets, asteroids, meteorites etc.

As it is believed that the two ingredients in the above recipe, the
presence of rotation and the magnetic field are common in GMCs, the
planetary systems may also be fairly common. In other words, most
stars in the process of formation should also pick up a few planets.

Until 1991, planets around stars other than the Sun (often referred
to as *extra-solar planets*) were not known. The situation changed dramatically in the last decade of the last century and we will briefly review
the current status of extra-solar planets later in this book.

Two questions

Although we have given a plausible account of how stars form, there are
two gaps in our discussion given so far; one in the beginning and one in
the end.

We started with the assumption that a GMC has parts that contract
and are therefore denser than the rest of the diffuse material in the cloud.
How did this come about in the first place? In the initially diffuse state
the cloud's own gravity is too weak to set off contraction. There has to
be some initial external push given to parts of the GMC to set them
along the path of contraction. Once a part begins to contract, gravity
can take over and accelerate the process. What was this external cause

that started it all? We will discuss a probable answer to the question in Chapter 8.

The question at the end is, of course, that of identifying the mysterious source of energy that comes into operation once the central part of the star is hot enough. As mentioned before, this epoch marks the beginning of the star's life as a 'full-grown adult'. We will take up this question next.

Chapter 6

The Secret of Stellar Energy

By the mid 1920's astrophysicists had a reasonably accurate picture of what a typical star is made of. Regardless of the considerations of the previous chapter which give a scenario about *how* a star is born, much of which came to be understood later, the astrophysicists of the 1920's took the star as a ready-made object and investigated the conditions under which it could maintain its shape and appearance for hundreds of millions of years. Pioneering work in this field was done by Sir Arthur Stanley Eddington (1882–1944) at Cambridge. Even today astrophysicists working on stellar interiors start with the equations set up by Eddington. His book *The Internal Constitution of the Stars* continues to be a classic on this subject.

One major advance in the study of stars that enabled Eddington to set up his equations of stellar structure came from the work of the Indian astrophysicist Meghnad Saha (see Figure 6.1). Known as the Saha equation, this study is concerned with the composition of hot plasma in thermodynamic equilibrium. If we imagine an enclosure containing gas being slowly heated, as its temperature rises, so will the speed of the atoms: the randomly moving atoms will move faster and faster. As and when two atoms collide, the energy of collision may break one of the atoms, releasing one or more trapped electrons. In short the atom gets ionized and the number of ionized atoms in relation to bound neutral

Figure 6.1: Meghnad Saha (1893–1956).

atoms will increase. Saha's equation tells us how this ratio can be determined for a given system at high temperature. Spectroscopists can, by looking at the star's spectrum and using Saha's equation, determine the above ratio and hence the surface temperature of the star. And knowing this important information enabled theoretical efforts, like those by Eddington, to determine the physical conditions *within* the star.

Without going into peripheral details we begin with the ideas behind Eddington's equations.

The equation of support

Let us start with the assumption that the star is a ball of hot plasma, which is somehow able to maintain its shape and size. The word 'plasma' indicates that we anticipate that the temperature of the ball of gas is so

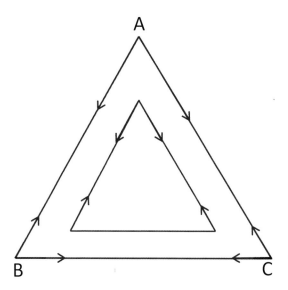

Figure 6.2: The triangle ABC shrinks inwards as the masses at A, B and C move towards each other attracted via the law of gravitation.

high that the atoms in it get stripped of their outer electrons. We now consider the crucial issue of how such a ball of plasma maintains its size and shape.

In Figure 6.2 we have three points A, B, C forming an equilateral triangle in the star's interior. Suppose that at these points we have equal masses and suppose further that no other matter exists in the star except these three masses at A, B, C.

According to Newton's law of gravitation, these masses will attract one another. If there is nothing to hold them in their positions, they will move towards one another with the result that the triangle ABC will shrink.

The above simple example illustrates the general tendency of all other points in the star which we so far ignored. They all attract one another and so the star as a whole tends to shrink. In fact this tendency to shrink is no different from that present in the primeval molecular cloud of the last chapter. Except that in this case the star is somehow prevented from shrinking.

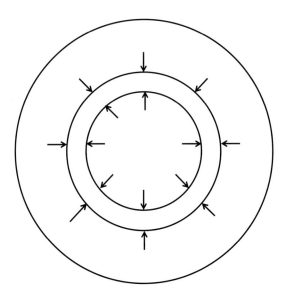

Figure 6.3: In the star, the inner spherical surface has higher pressure than the outer one.

The opposing force that holds the star in its extended state is, of course, that coming from the pressure in its own gas. This pressure has to be maximum at the centre of the star, decreasing steadily outwards until it falls to zero value at the star's surface. Figure 6.3 illustrates how such an outward pressure helps support the star.

In Figure 6.3 we have a spherical shell of gas concentric with the centre of the star. As we mentioned earlier, this shell has a tendency to contract. But . . . , now we see how the pressure in the gas produces opposing forces on the shell's inner and outer surfaces. To understand the situation better, consider the pressure felt by a deep sea diver. As he goes deeper, he has to bear the weight of water above him which adds to the pressure of the atmosphere we all feel at the sea level. As shown by arrows in Figure 6.3, the force on the inner surface tends to press the shell outwards while the force on the outer surface presses it inwards. Since as in the case of the sea diver, the inner surface has greater pressure to reckon with than the outer one the former wins out. And so the pressure forces on the shell tend to blow it outwards.

This is an important issue and at the cost of repetition we note that this increase of pressure inwards is no different from what the deep sea diver feels as he goes deeper and deeper below the sea level. Such a diver has to bear the weight of the column of water above him, besides the pressure of the atmosphere at the surface. As he dives deeper, the height of the column above him increases and thus adds to the pressure. In a spherical star the same effect makes the weight of the outer layers add to the pressure inside and this grows as one goes towards the centre of the star.

Thus we have the opposing force arising from pressure which has to match the inward pull of *gravity exactly*. If there is the slightest imbalance, the star will either blow outwards or contract inwards. Indeed, as we shall see later, such imbalances do occur in a star's life on certain occasions. Imbalances can be so large that the star can *explode* and lose a good deal of matter in the process or it can *implode* to a very condensed form.

As an example of the matter let us consider a thought experiment — that is, imagine a situation that we cannot, of course, realize in practise. Imagine that by magic the Sun instantly finds itself without any pressure: it will then begin to contract under the force of gravity. Calculations show that this force grows in strength as the star shrinks and the contraction will proceed with a rapidly rising tempo until the whole Sun shrinks to a point. An observer sitting on the Sun will find by his watch that the entire operation takes no more than twenty-nine minutes!

This example, although somewhat extreme, illustrates the importance of exact balance between pressure force and gravity. In Chapter 10 we will recall this extreme example in a different context where it will appear quite natural.

The temperature within the star

Having realized that the star must have large pressures in its interior we now proceed further to find its consequences. What can the pressures be due to?

There are two ways in which a star can have huge pressures inside. The first way which is more obvious is the pressure existing in any gas

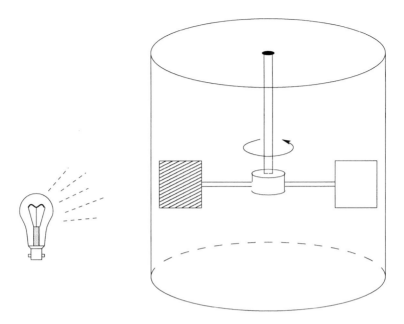

Figure 6.4: In the toy illustrated above each of the panels A and B have a reflecting face (shown on B) and an absorbing face (shown shaded on A). Light is absorbed by the latter and reflected by the former, thus imparting a net push on each panel to set the pair rotating.

which is not absolutely cold. Indeed, we know from observations that the external surface of a star has a temperature of several thousand degrees.

If we make reasonable assumptions based on the studies of gases or plasma heated to high temperatures in the laboratory, we come to the conclusion that just as the pressure in the star increases as we go inwards, so does the temperature. Thus the temperature may start at several thousand degrees at the star's surface and increase steadily inwards till it may exceed a million degrees at the centre.

The second way in which a star can produce high pressures inside is through radiation. The revolving toy shown in Figure 6.4 works on radiation pressure. There light falling on the panes is absorbed by the rough side and reflected by the bright side. The result is a net pressure which rotates the panes.

We are accustomed to the gas pressure coming from the Earth's atmosphere — a pressure that can hold up a 30 inch or 760 mm column of mercury vertically. We don't realize that the sunlight falling on the Earth also carries pressure from radiation — because it is very tiny compared to the gas pressure. But within a star, at high temperatures of hundreds of thousands to millions of degrees the pressure exerted by radiation is terrific. We recall from Chapter 2 that light is made of particles called photons which carry packets of energy. When a swarm of such photons of high density and energy impinges on a surface it exerts enormous pressure on it. And so radiation pressure also becomes important in many stars.

The transfer of radiation

Figure 6.5 shows that an intense source of energy generates radiation at the centre O of the star. As the photons try to push their way out they generate radiation pressure. But what happens eventually to these photons produced at the centre?

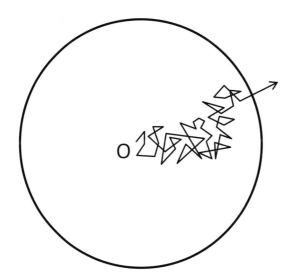

Figure 6.5: The jagged path of a photon from the centre on its way out.

A typical photon is absorbed by gas atoms in the star and then re-emitted in another direction. Of course if this happens a lot of times, the photon will find it very hard to make its way out of the star. It is estimated that a typical photon generated in the solar core is bumped around a hundred billion billion (10^{20}) times before it escapes from the surface. The process takes about 300 years! One may liken the situation to that of a person trying to make his way out of a crowded fair ... he is pushed around in different directions by the crowd and it takes a lot of effort and time for him to come out!

One of Eddington's equations deals with how the energy of radiation is progressively transferred from the interior of the star to its surface. It takes into account the *opacity* of the stellar material, that is, its ability to absorb and scatter the photons as they try to come out. Again, the calculations are based on what we know about opacity of hot gases from laboratory experiments and the theory of radiation interacting with a hot gas.

Thus Eddington had altogether a set of four equations which describe the different aspects of a star but which together are necessary if we wish to know how a star functions as a complete unit. There was, however, one snag in the whole picture, a snag that can be spotted in Figure 6.5.

In that figure it is *assumed* that there is source of radiation (the star's core) which pours out energy, energy that is eventually radiated outwards by the star and which also keeps it hot. But what is this mysterious source of energy? The fifth equation in Eddington's set dealt with this question.

From Kelvin and Helmholtz to Eddington

Two distinguished physicists of the nineteenth century had worried about this question and had offered what had then appeared to them to be a plausible solution. Lord Kelvin in Great Britain and Baron von Helmholtz in Germany had both suggested that the typical star derives its energy of radiation from its vast storehouse of gravitational energy.

The following example illustrates this idea. Imagine a spherical ball of matter of a fixed radius. Suppose we want to break it apart and carry all the pieces out to great distances. To achieve this goal we have to work against the gravitational attraction of the ball. Each piece that is taken away has the tendency to fall back on the ball and must therefore be pushed away with a force greater than the gravitational pull of the remnant. In other words, we have to call in an external agency to counter the ball's gravity and that agency will lose energy in the process of breaking up the ball and dispersing it to large distances.

Let us call the two states of the ball, state I and state II. State I is a tightly bound ball of radius R, say, while state II is the infinitely dispersed one. Energy must be spent to go from state I to state II.

One of the basic principles of physics, the so-called *law of conservation of energy* guarantees that the total store of energy remains unchanged in any physical process. Thus, the energy that was spent by the external agency in converting state I to state II is not "lost"; but it gets "stored" in state II. State II will therefore have higher energy than state I. The difference of energy is manifested as *gravitational potential energy*.

It is a good convention to set the "zero" level of energy equal to that stored in the condition of state II; for this is the state when all bits and pieces are so far apart from one another that they hardly "feel" the gravitational attraction any more. But then, from the above arguments, state I being less energetic than state II must have *negative* energy. A calculation using Newton's law of gravitation and based on the assumption that in state I the ball has uniform density, leads to the result that the gravitational potential energy is given by

$$E = -\frac{3}{5}\frac{GM^2}{R}.$$

Here M is the mass of the ball and G the constant of gravitation.

Figure 6.6 illustrates how E changes as the ball of matter slowly contracts, keeping its density uniform all the time. As R decreases, E becomes more and more negative. In other words the contracting ball loses its store of energy.

Kelvin and Helmholtz argued that this is what happens to stars like the Sun. Although a star is by no means a uniformly dense ball of matter, the above arguments apply to it with minor changes. In particular the

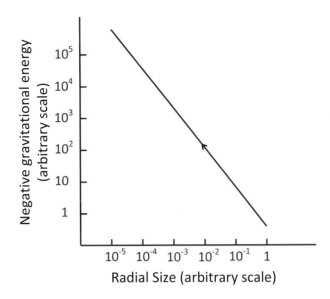

Figure 6.6: The negative of the gravitational energy E is plotted against the radial size R of a collapsing object. The scale is logarithmic. Thus a reduction in size hundred-fold makes $-E$ go up by the factor 100.

coefficient $3/5$ in the above formula changes to another fraction. We will ignore these matters of detail in our discussion and work with the above example of the uniformly dense ball.

So according to Kelvin and Helmholtz a star slowly contracts and loses energy which appears as its radiation. If we take the specific example of the Sun, we can calculate the energy it has lost since its contraction from an infinitely dispersed cloud of gas (state II of our example) to its present state of radius of about 700 million metres (state I). The Sun's mass is 2×10^{30} kg. So the above formula for the uniform ball gives the energy lost through contraction as

$$E_\odot = 2.4 \times 10^{41} \text{ joules.}$$

(A one-watt light bulb consumes 1 joule of energy per second.)

This figure appears enormous, but let us put it in perspective with the Sun's rate of radiation of energy, which is at present

$$L_\odot = 4 \times 10^{26} \text{ watts.}$$

Assuming that the Sun has been steadily radiating at this rate, it would have used up the total amount of energy $E_\odot$ in a matter of

$$\frac{E_\odot}{L_\odot} = \frac{2.4 \times 10^{41}}{4 \times 10^{26}} \text{ seconds} = 20 \text{ million years.}$$

By human time scales this is a comfortably long time span. But not by geophysical standards! We have independent estimates of the age of the Earth from the estimates of its rocks and minerals, as calculated by geophysicists. The geophysical estimate of the age of the Earth and the Solar System is around 4.6 billion years and for most of this time the Sun must have been shining at a rate not too different from its present rate. For, the fossil data on the Earth suggests the existence of primitive life on it at least 3 billion years ago; and life is believed to be intimately connected with steady supply of energy from the Sun. If, as given by the Kelvin-Helmholtz contraction hypothesis, the Sun has been shining only for a few million years, it would be hard to explain the much longer time scales from geophysics.

By the mid-1920's it was realized that the Kelvin-Helmholtz hypothesis was not the right answer to the ultimate source of stellar energy. Some entirely new and considerably more powerful source was required.

This is where Eddington entered into the picture with his idea of *thermonuclear fusion*. Taking seriously the suggestion first made by Jean Baptiste Perrin that energy would be released if four nuclei of hydrogen were somehow converted into a nucleus of helium, Eddington argued that the key to stellar energy lay not in the gravitational potential energy but in the energy locked inside atomic nuclei. We have already seen how at temperature of the order of a few thousand degrees an atom cannot retain its whole structure intact but gets stripped off of its outer electrons and becomes ionized. At these temperatures the nucleus of an atom remains intact since it is much more tightly bound than the atom as a whole is. Eddington felt that at the temperature of *millions of degrees* operating in the star's centre, however, even the tightly bound nucleus begins to be affected.

The mid-1920's had barely seen the beginning of the subject of atomic physics tackled according to the newly discovered rules of quantum theory. Little, if anything was known about how atomic nuclei behave. Eddington's arguments were therefore based largely on conjecture

and intuition. Still the idea that atomic nuclei could be broken apart or
fused together was considered so radical at that time that atomic physicists refused to buy it, not even at such high temperatures as those which
Eddington's calculations implied for stellar cores. Eddington nevertheless was confident that only herein lay the answer to the long standing
question: Why do stars shine?

Writing in his classic book, *The Internal Constitution of the Stars*,
mentioned earlier Eddington had this to say to the doubting Thomases:
"We do not argue with the critic who tells us that stars are not hot
enough for this purpose. We tell him to go find a hotter place."

The star as a nuclear reactor

Within two decades Eddington was vindicated. By the end of the 1930's
several developments had taken place in the field of nuclear physics that
made it possible not only to visualize how nuclei behave at very high
temperatures but also to make detailed calculations about how much
energy is available from the locked-up nuclear store that Eddington
wanted opened. Let us look at the problem from the modern standpoint.

Figure 6.7 shows two nuclei, one of hydrogen, the other of helium.
The hydrogen nucleus contains just one particle, the positively charged
particle called the *proton*. The helium nucleus is more complex; it has
four particles, two of which are protons while the remaining two are
the electrically neutral particles called the *neutrons*. Let us denote the
proton by the letter p and the neutron by the letter n.

The first important thing to note is that in the helium nucleus
the two protons are happily staying close together. To someone who
has studied only electrostatics, this result may appear strange. For,
Coulomb's law of electrostatic repulsion states that like charges repel
and that too with a force rising inversely as the square of their distance
apart. So the two protons contained within a nuclear distance of 10^{-15}
metre must be experiencing an enormous force of repulsion. How then
do they manage to stick together?

They manage to do so because, at such short distances, a new force
comes into play. This force is one of attraction and is known as the
nuclear binding force. As stated earlier, it is much stronger at nuclear

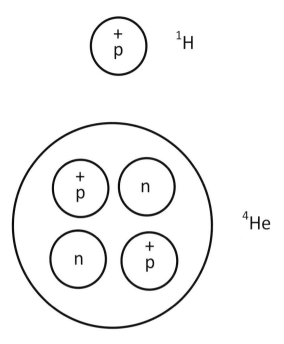

Figure 6.7: The hydrogen and helium nuclei.

distances than the electrical force. Also it acts uniformly on neutrons and protons, that is, it does not depend on whether the particle carries an electric charge or not.

The nuclear force however does not extend over long distances, over distances beyond the nuclear size of round 10^{-15} metre. So if we want to build a helium nucleus out of four protons we have to bring these particles close enough to counter the electric repulsion between the protons.

Imagine a situation in which we arrange to hurl two protons towards each other. If we don't throw them fast enough, they will slow down to a halt outside the zone of their nuclear attraction and move away again because of their mutual electrical repulsion. However, if we manage to throw them towards each other very fast, they may come so close that they find themselves within the zone of influence of the nuclear binding force. In that case they continue to remain together. We have then achieved a *fusion of nuclei*.

Clearly for fusion to be possible we want the protons to move very fast. In a very hot gas the protons do move very fast, though in random directions. In fact the temperature of the gas is a measure of how fast the average particle in it moves. It is possible, provided the temperature is high enough, for two randomly moving protons to come so close as to get fused. What is critical temperature for this purpose?

It turned out that the temperatures calculated by Eddington for the centres of stars, in the range of ten to forty million degrees *were* high enough to achieve the fusion process. But something more was needed: that in the process some energy should be released.

A look at Figure 6.7 shows that if we bring four nuclei of hydrogen together we do not quite make a helium nucleus. Two of the protons have somehow got to be replaced by neutrons. *In the fusion reaction this does happen.* Symbolically we write the reaction as follows:

$$4\,{}^1\text{H} \rightarrow {}^4\text{He} + 2e^+ + 2\nu + \text{ energy.}$$

The superscripts 1 and 4 indicate the number of particles in the nuclei of hydrogen (H) and helium (He) respectively. e^+ is a positively charged particle called the *positron* while ν is a neutral particle called the *neutrino*. Thus of the four units of charges in the original hydrogen nuclei, two go into the helium nucleus and two are carried by the positrons. Of the four participating protons two get converted to neutrons. The positron, by the way, is the antimatter counterpart of the electron. It has the opposite charge but the same mass as of the electron. If brought together the electron and the positron annihilate each other producing pure radiation.

The relevant feature of the fusion reaction is the last term in it which tells us that energy *is* released in the fusion process. Where does it come from? If we calculate the total mass of the four hydrogen nuclei that participated in the reaction and compare it with the mass of the resulting helium nucleus we find that the former was somewhat higher. The positrons and neutrinos are very light particles and they do not make up for the mass loss.

In other words, the law of conservation of mass appears to be violated. However, this need not be a source of worry to the twentieth-century physicist. For, according to Einstein's famous relation,

$$E = Mc^2,$$

the mass loss M is made up as energy whose magnitude is obtained by multiplying M by the square of the velocity of light.

What fraction of mass of the hydrogen that was so fused gets converted to energy in this way? Only 7 parts in 1000. Small though this fraction is, it is sufficient to provide the energy required to keep a star like the Sun shining for billions of years.

Let us look at the problem from the terrestrial point of view. Suppose we have a fusion reactor which is able to fuse together 1 kilogram of hydrogen into helium. How much energy will it generate? Using our result above, we find that of the 1 kg fuel, only 7 grams will be effectively converted into energy. The Einstein formula gives, however, the incredible result that the amount of energy generated is 7×10^{14} joules. This is the energy output of a megawatt generator running continuously over a period of about 20 years! The stellar models calculated by the nuclear physicist Hans Bethe in the late 1930s showed how everything fits, and we get the answer to the age old question: Why do stars shine? For this work Bethe was awarded a Nobel Prize.

This example illustrates the tremendous potential of fusion reactors once terrestrial technology advances far enough to produce them. Technology to date has succeeded in producing the hydrogen bomb. The difference between the fusion reactor and the bomb is that while both use the same fusion reaction, the former generates energy in a controlled and steady rate while the latter generates it explosively.

Stars are able to achieve controlled nuclear fusion because of the gravity-induced high pressures in their central regions. Human technology has to find other means of achieving controlled fusion, because it cannot emulate the stellar scenario: it lacks the gigantic force of gravity available to the stars. The present efforts try to use magnetic confinement of plasma instead.

Chapter 7

The Origin of Chemical Elements

In the previous chapter we found that the source of a star's radiant energy lies at its centre and that it has the form of a nuclear fusion reactor. We will now see how, as the star ages, this nuclear reactor also changes and this change is of importance for two reasons. Firstly, as we will see, if the nuclear energy source changes, this affects the physical state of the star including its outward appearance. Thus we will be able to understand the location of stars on the H-R diagram *not* on the main sequence. Secondly, the range of nuclear reactions that go on in the core region of the star give rise to nuclei of various chemical elements. Thus our attempt to work out the details of stellar evolution gives us a bonus in telling us how the elements we see around us came into existence. In this chapter we will be concerned with both of these aspects.

Stars on the main sequence

The simplest nuclear reactor is the one we discussed in the last chapter. Its fuel consists of the nuclei of hydrogen, which are simply protons. The four protons are brought together to form a nucleus of the helium atom. And in this fusion process energy is released.

Before this process was known, Eddington's equations of stellar structure were incomplete in the sense that one equation from the complete set was missing. In such a situation, the best that the astrophysicist could do was to calculate the star's luminosity L, given two of its basic parameters: its mass M and radius R. When the nuclear energy generation problem was resolved, it became possible to do better: to calculate the luminosity L as well as the radius R in terms of the single parameter, the star's mass M. Since it is easier for the astronomer to know the star's luminosity than its radius, the mass-luminosity relation became the key result to come out of these calculations, a result that is of direct relevance to observations. We will illustrate this point with the help of the H-R diagram.

There are two crucial ways in which the mass of a star affects its overall internal constitution. The first of these determines how energy is transported from the star's deep interior to its outer surface while the second relates to the route by which the nuclear reactions proceed from hydrogen to helium. We will look at both these issues in that order.

Our daily experience tells us that in general there are *three* ways in which heat energy can be transported from one point to another. In *conduction* heat is passed on from neighbour to neighbour by atoms and the molecules of the heated material, while these particles remain fixed in place. This process operates, for example, in solid metals. In *convection*, common in liquids and gases, small units of the heated substance bodily move from one end to the other taking heat with them. This movement can be seen when water is heated to boiling point, for example. The third process is that of *radiation* where energy is transported by light photons.

The three processes can be compared to the following example. Suppose we are on a construction site and the work requires transporting bricks from one point A to another point B. One method is to have a row of labourers lined up from A to B. Each labourer passes on a brick received from his left neighbour, say, to his right neighbour. This is conduction. A second method is to employ workers to carry bricks from A to B, each one making the full trip between these two points. This is convection. The third method, likened to radiation, involves throwing bricks from A to B.

We have already seen in Chapter 5 how the last two processes dominate at different stages in the process of star formation. These two pro-

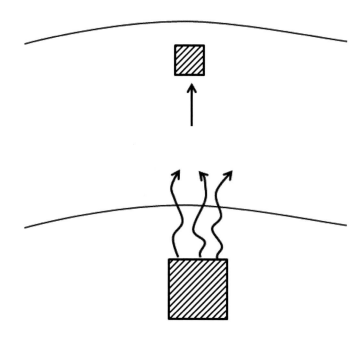

Figure 7.1: When the element of gaseous material shown by the hatched block (in the upper part of the figure) receives heat from inside it expands. It then has a tendency to float outwards as shown by curved arrows in the lower part of the figure. The conditions in the outer layers will determine if this element will be permitted to come out.

cesses also take place in a fully formed star: again, not with equal levels of efficiency. It may happen that in some parts of the star conditions are more conducive for gas particles to move bodily and convect heat from the inner region to the outer region. This can happen when, on receiving heat from the central source, a unit of gas particles expands and becomes lighter compared to its surroundings so that it floats outwards as shown in Figure 7.1. In some cases this may not be the best way of energy transport and radiation (in the form of light photons) is the more efficient mechanism. But, as we saw in the previous chapter, the opacity of stellar material may, in some cases, make it difficult for a light photon to come out.

Thus each process has its *pros* and *cons*. If the opacity is high enough there is a steep decline of temperature in the outward direction and

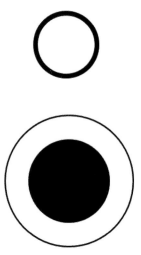

Figure 7.2: Energy transport in stars. The shaded region denotes convective transport of energy; the unshaded region is where energy transport is radiative. The star above is less massive and the one below is more massive than the Sun.

convection may dominate while radiation would be important if the opacity is not too high. Quantitative considerations (which go a lot farther than our qualitative arguments here) show that the mass of the star determines in what zone which process dominates.

Basically there are two zones, an inner core and an outer envelope (see Figure 7.2). In a very massive star the core is convective and the envelope radiative while in a low mass star these roles are reversed. The crucial mass limit where the transition takes place is around the Sun's mass $M_\odot$. The Sun itself belongs to the latter type (like a low mass star).

There is another important effect of the star's mass on the way the thermonuclear fusion takes place at its centre. The conversion of hydrogen to helium does not take place via a single reaction. Rather, there is a series of reactions which step by step add a proton or a neutron to the hydrogen nucleus until it has the composition of a helium nucleus. The reason for this is obvious. To get all four hydrogen nuclei in the same place at the same time is extremely hard. Rather one looks for step by step build up of these nuclei through two-body collisions. There are two routes by which this fusion of hydrogen to helium can be achieved. In low mass stars the more efficient route is the one called the *p–p chain*. It proceeds via the following set of reactions:

$$p + p \rightarrow {}^2\text{H} + e^+ + \nu,$$

$$^2\text{H} + p \rightarrow {}^3\text{He} + \text{ radiation},$$

$$^3\text{He} + {}^3\text{He} \rightarrow {}^4\text{He} + 2p.$$

Here p is the proton which is essentially the hydrogen nucleus. e^+ is the positron while ν is the neutrino which will occupy our attention later. ^2H is the nucleus of 'heavy' hydrogen (which contains an extra neutron) while ^3He is a lighter nucleus of helium (containing one fewer neutron). The final product is the standard ^4He nucleus of helium. As mentioned earlier, the series of nuclear reactions above proceed by successive addition of protons, which is why the process is designated the *p–p* chain. Notice that the eventual outcome of this chain is that four protons come together and get converted to the helium nucleus.

In high mass stars this process is not a very efficient one and is dominated by another called the *CNO cycle* in which the nuclei of carbon, nitrogen and oxygen play the roles of *catalysts*. The role of a catalyst in a chemical or a nuclear process is that of a mediator to speed up its operation; at the end of a process the participants of the process undergo changes but the catalyst itself is left intact. This is seen to happen in the *CNO* cycle reactions given below:

$$^{12}\text{C} + p \rightarrow {}^{13}\text{N} + \text{ radiation},$$

$$^{13}\text{C} \rightarrow {}^{13}\text{C} + e^+ + \nu,$$

$$^{13}\text{C} + p \rightarrow {}^{14}\text{N} + \text{ radiation},$$

$$^{14}\text{N} + p \rightarrow {}^{15}\text{O} + \text{ radiation},$$

$$^{15}\text{O} \rightarrow {}^{15}\text{N} + e^+ + \nu,$$

$$^{15}\text{N} + p \rightarrow {}^4\text{He} + {}^{12}\text{C}.$$

The nuclei of C, N, O (carbon, nitrogen, oxygen), however, have to be around in small quantities, especially ^{12}C, to make this process work.[1]

---

[1]Thus the *CNO* process presupposes the existence in the star of elements heavier than ^1H and ^4He. How did these elements get there in the first place? We will return to this point in the next chapter.

At the end of the cycle the C, N, O content of the star is left intact, however. The critical dividing line between high mass and low mass stars is not a sharp one but it is again somewhere around $M_\odot$. Thus the Sun and stars less massive than it process the hydrogen fuel by the p–p chain while more massive stars rely primarily on the CNO cycle.

All the component reactions in both the p–p chain and the CNO cycle, do not proceed at the same rate. In fact in both processes the first reaction is the slowest and determines the overall rate of fusion. The characteristic time runs from several billion years for very low mass stars to tens or hundreds of millions of years for very massive ones.

We will subsequently find that after the synthesis of helium is complete the star moves on to process heavier nuclei in its core. These later processes are, however, considerably faster than the process of helium synthesis. Thus, for most of its life the star will be making helium in its interior. Looking back at the H-R diagram (Figure 7.3) we see that this is the reason why the main sequence contains the largest number of points; for stars on the main sequence are the ones operating the H $\rightarrow$ He fusion reactor slowly but steadily for a long time.

The mass-luminosity relation for the main sequence stars takes the form

$$L \propto M^n$$

where $n = 1.6$ for the low mass stars $(M \lesssim M_\odot)$ and $n = 5.4$ for high mass stars $(M \gtrsim M_\odot)$.

Thus as we go "up" the main sequence from the end B and to end A we encounter stars of higher and higher mass and greater and greater luminosity. We can also compute the surface temperatures of the star models of different mass and check whether the surface temperature so computed agrees with that in the H-R diagram. Such a calculation provides a check on the overall correctness of the theory.

It is a measure of the success of the present-day calculations that very good agreement is found between theory and observations. Starting with the equations set up by Eddington, the modern-day astrophysicist uses the best available data from atomic and nuclear physics to perform computer calculations of the various differential equations. Indeed, much of the intricate theoretical details of stellar models could not be handled before the advent of fast electronic computers.

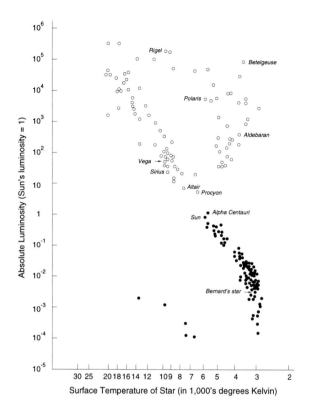

Figure 7.3: The H-R diagram reproduced from Figure 3.4.

There is another *experimental* check that we can make on these models: viz., the detection of neutrinos emitted in the thermonuclear reactions. The detection and study of such neutrinos is technically possible and forms an important line of investigations. We will postpone its discussion, however, to the final chapter of this book.

The red giant stars

Large though the nuclear energy source is within the Sun, it is finite and will sooner or later be exhausted. This can happen in the way illustrated by Figure 7.4. We have here a central core which was originally all hydrogen but has now become all helium, thanks to the working of the

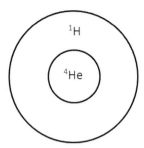

Figure 7.4: At the end of the main sequence phase the star has a helium core and an envelope of hydrogen.

thermonuclear reactor. We recall that for the reactor to function the temperature must be high enough. We also recall that the temperature in a star drops outwards.

So, in Figure 7.4 we have the following situation. Within the core the temperature is high enough for the reactor to function, but there is no fuel left to process. Outside the core there is plenty of hydrogen fuel but it is not hot enough to take part in the fusion process. Thus the star appears to have reached a dead end so far as its active life is concerned. For the Sun this will happen when it has converted 12 percent of its hydrogen to helium. However, this impasse is of a temporary nature and resolves itself in the following way.

Recall from Chapter 6 that energy generation in the centre of the star is responsible for providing the high temperatures and pressures to counteract the force of gravity. If the source of energy is switched off, the pressures can no longer be maintained. Although the effect of the resulting imbalance is felt all over the star, it is naturally most drastic in the central core of helium on which the weight of the outer envelope is tremendous. The pressures which supported this weight are no longer adequate. Unable to prevent the contracting tendency of the star's gravity the core begins to shrink. However, this is not the end of the story for the star! On the contrary, this shrinkage opens up interesting new possibilities for the star's future.

As the helium core contracts it heats up as any gas would if it were compressed. Drawing the analogy with what happened earlier to hydrogen, we now ask the question: "Is it possible for helium to be heated to a high enough temperature so that it too becomes active fuel for another fusion process?" For, if it were possible, then the star's active life is not yet over! It could keep generating energy from a different fusion process.

The question remained unanswered till the mid 1950's. Nuclear experimental data showed that as we ascend the mass ladder towards heavier nuclei, those immediately following ^{4}He are not stable. For example, we may try to fuse two helium nuclei together to form a nucleus of beryllium ^{8}Be. The nucleus ^{8}Be, however, breaks apart as soon as it is formed! So the fusion process could not go on along this route. Similarly there is no stable nucleus of five particles that could be formed by adding a neutron or a proton to ^{4}He. Imagine that you are climbing a ladder when you discover that two successive rungs are weak and cannot be used for stepping on. At this stage, what would you do? You may be more adventurous and try to step on the next new rung which may be strong. Hopping up three rungs will, however, be difficult.

Nuclear physicists like Ed Saltpeter tried such an option. The possibility of three helium nuclei fusing together produces carbon which is a stable nucleus. So why not try this route? However, this had one problem. A random collision may bring two helium nuclei together fairly often. But a triple collision in which three helium nuclei arrive at the same place at the same time is very rare, with the result that calculations yield very little energy in this process. In short, as a source of stellar energy this route seemed doomed.

At this stage the suggestion that led to a breakthrough came not from a nuclear physicist but from a theoretical astrophysicist. Fred Hoyle, who made the suggestion did so because he felt convinced that since we see stars of different types besides those on the main sequence, there had to be an effective way of combining helium nuclei whereby a star manages to keep its fusion reactor going beyond the main sequence. Also, Hoyle felt that the Saltpeter idea was correct in the sense that it yielded carbon which is an important element in the universe. Such a reaction would be needed to make carbon in the universe. But what can be done to get round the very slow rate of this reaction? Hoyle proposed that the three helium nuclei fuse together in a *resonant* reaction to form a nucleus of carbon, *in an excited state*:

$$3\,^4\text{He} \rightarrow {}^{12}\text{C}^\star.$$

The asterisk on C denotes the fact that the carbon nucleus is excited, that is, it has more energy than the ordinary nucleus. In this state the nucleus cannot, however, remain for very long; it must decay to the ordinary state by releasing the extra energy:

$$^{12}C^\star \rightarrow {}^{12}C + \text{ radiation.}$$

What is a resonant reaction? How does this process help? Hoyle argued that to compensate for the rarity of the three-body collision event, the resulting fusion reaction must proceed very swiftly as if it were highly favoured to happen. Physicists call such favoured processes *resonant* processes. Figure 7.5 illustrates how a simple oscillating pendulum resonates when its bob is hit periodically. For the resonance to occur the frequency with which the bob is hit must equal the natural frequency of the pendulum. In a musical instrument like the violin

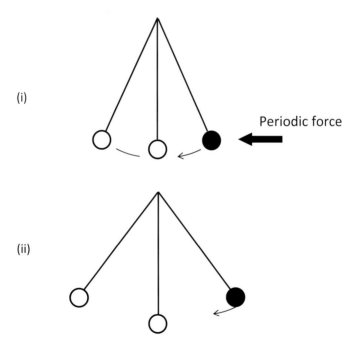

Figure 7.5: The application of a periodic force increases the amplitude of oscillation of a simple pendulum from initially small values in (i) to large values in (ii) *provided* the period of the force matches that of the pendulum.

the string frequency is matched by the frequency of air trapped in the instrument's enclosure to produce an augmentation of volume.

Likewise, to achieve resonance in the fusion reaction Hoyle argued that the excited carbon nucleus had to have energy *exactly equal to the combined energy of the three helium nuclei.* His calculation led to the conclusion that the above match could take place if the carbon nucleus had higher energy than normal. So he suggested to the nuclear physicists to look for such a resonant excited state of carbon in laboratory experiments. Nuclear physicists were skeptical of the prediction but they looked for it. Hoyle's nuclear physicist colleagues at the California Institute of Technology led by Ward Whaling experimentally confirmed the existence of an excited carbon state with energy exactly as predicted! Here we have an excellent example of astronomical considerations leading to a discovery in fundamental science.

In this context it is worth noting that human bodies contain 65 percent oxygen and 18 percent carbon (the remaining being mostly hydrogen). If, elements like carbon and oxygen are to be available in the universe to sustain life as on the Earth, then these have to be made somewhere and stars provide a plausible locale. But if they are to be made in stars then there has to be some way of ascending the nucleosynthesis ladder above ^{4}He. This was another reason that motivated Hoyle to look for this way of synthesizing carbon. It is curious, as Hoyle pointed out, that our very existence seems to depend on the rather subtle effect of there being an appropriate energy level in the carbon nucleus — the excited level mentioned above!

The temperatures at which the three helium nuclei fuse into a carbon nucleus lie in the range of 100–200 million degrees. The fusion process therefore starts when these temperatures are attained in the shrinking core. Since the helium nucleus (when it was first discovered in the laboratory experiments on radioactivity) was called the α-particle, the above reaction is sometimes called the *triple-alpha* process.

The production of energy by fusion generates high temperatures and pressures which therefore stop any further contraction of the core. However, the pressure structure in the entire star has to adjust itself to this new situation. Remembering that the pressure at the surface of the star has to be zero, the readjustment that is needed blows up the

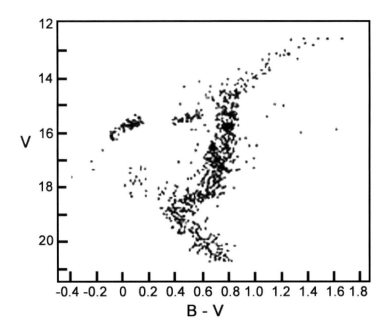

Figure 7.6: The H-R diagram showing the branching off of the group of stars in the globular cluster M3 along the giant branch. (The upper branch is called the horizontal branch which we have not discussed in the text.)

outer envelope to a size considerably larger than before. The star thus becomes a "giant".

When the Sun reaches this stage it will have grown so distended that its outer surface would gobble up the inner planets including the Earth. And as the envelope expands the star also cools down. Thus the outer surface of a giant star has a considerably lower temperature than that of a main sequence star. And, of course, the colour-temperature relationship discussed in Chapter 3 implies that the star will look *reddish*.

Figure 7.6 shows that the H-R diagram of stars in the globular cluster. We see the main sequence stars as well as the stars branching off it towards the right, as they become red giants. The general rule is that the more massive a star, the faster does it evolve off the main sequence. For, with increased mass the central temperature is higher and the nuclear fuel is processed faster.

The alpha process

As the reader will have guessed, the helium fuel will also come to an end sooner or later, and the star will be back to the situation it had found itself in when the hydrogen fuel was exhausted. As happened earlier, if the nuclear energy source is apparently exhausted, the core will start shrinking and heating up.

And as before, the star finds a new nuclear reaction to generate energy using the same trick as before: raise the inner core's temperature by contraction until a new fusion reaction can take place. Thus the next reaction adds another α-particle to the already formed ^{12}C, to make a nucleus of oxygen:

$$^{12}C + {}^4He \rightarrow {}^{16}O.$$

This reaction becomes possible at temperatures above 200 million degrees. When its fuel is exhausted, again the same sequence of events would take place, viz. the core contraction and heating up followed by another fusion reaction in which the alpha particle is added to the oxygen nucleus to form neon. At still higher temperatures even heavier nuclei are formed by successive additions of the helium nuclei. Thus we get nuclei in the series with atomic masses increasing in steps of 4:

$$^{16}O, \ {}^{20}Ne, \ {}^{24}Mg, \ {}^{28}Si, \ {}^{32}S, \ldots$$

these being oxygen, neon, magnesium, silicon, sulphur etc. This ladder-like sequence is called the *alpha ladder* and the nuclear process causing it the *alpha process*. Eventually, two silicon nuclei are fused to form a nucleus of nickel at a temperature that has by now risen to 3.5 billion degrees:

$$^{28}Si + {}^{28}Si \rightarrow {}^{56}Ni.$$

For reasons which we shall explain shortly, the fusion process terminates just above here. Beyond the nuclei of the "iron group", namely, iron, cobalt and nickel, the star's fusion reactors cannot proceed.

By this time the star has grown to full giant shape because whenever it exhausted a particular fuel its core contracted till the triggering off of a new fusion reaction and then its envelope expanded. Although, the star at the beginning may have had a homogeneous predominantly

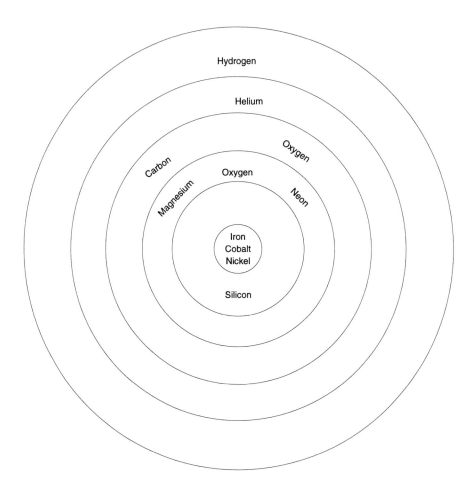

Figure 7.7: The interior of a red giant star simulates an onion-skin like distribution with shells of heavier nuclei towards the centre.

hydrogenic composition, it has now developed a layered onion skin-like structure. Shown in Figure 7.7, this structure has the heaviest elements (the iron group) at the centre with progressively lighter elements as we move outwards to cooler regions. The outermost parts will still be predominantly hydrogenic since they are too cool to take part in any nuclear fusion.

The end of the road

Let us leave the stars for a while and go back to the unanswered question as to why nuclear fusion stops at the iron group.

To seek the answer we recall the two forces of nature that act between the nuclear particles. The electric force of repulsion acts between two protons but leaves the neutrons immune. The nuclear force of attraction acts equally strongly on all protons and neutrons. For a small nucleus like ^{4}He, the latter force is far stronger than the former and the nucleus is tightly bound. However, if we have a very heavy nucleus, say having more than 50 particles, its size is large too. And at large distances the nuclear binding force does not work as effectively. Also, a large nucleus has a large number of protons whose electric repulsion will be considerable. Thus beyond a certain limit of mass, the nucleus is not as tightly bound as the ones lighter than itself. The nuclei of the iron group are the most tightly bound nuclei. By adding more protons or neutrons we make new nuclei that are less tightly bound than those of the iron group.

Figure 7.8 illustrates this point graphically. It shows the "binding energy per nucleon" plotted against the different nuclei. A nucleon is either a proton or a neutron and the binding energy is the amount of energy that must be exerted to strip the nucleon out from the nucleus. As we see in Figure 7.8 the binding energy per nucleon is higher for helium than for hydrogen. That means that energy must be put in to break apart the helium nucleus to make it into hydrogen nuclei. Conversely, if we put hydrogen nuclei together we get helium nuclei and some energy. This is the reason why stars get energy by the fusion process. Also, the fusion process will work for heavier nuclei so long as we keep *ascending* the binding energy curve. Once we reach the peak (at the iron group) we can only descend. And so fusion will not help any more.

In fact it is clear from Figure 7.8 that the bulk of the fusion energy is released in the first step of the nucleosynthesis, from the conversion of hydrogen to helium. For the rest of the reactions the energy released is much less. Nor does later fusion greatly prolong a star's active life as red giant. Table 7.1 below gives the times spent by a massive star in processing the different fuels starting from hydrogen.

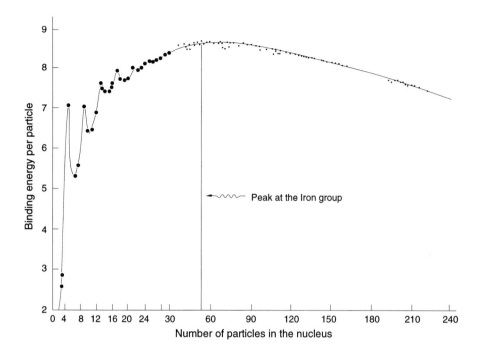

Figure 7.8: The plot of binding energy per nucleon shows that not all the nuclei are equally stable. The curve peaks at the iron group nuclei iron, cobalt, and nickel which are the stablest. (The energy is expressed in the unit mega electron volt that equals 1.6×10^{-13} joules.)

Table 7.1: Fusion times for different fuels in a star of $25M_\odot$.

Fuel	Temperature (in million degrees)	Time
hydrogen	60	7 million years
helium	240	500,000 years
carbon	930	600 years
neon	1750	1 year
oxygen	2300	6 months
silicon	4100	1 day

For a less massive star the times are much longer. Thus for the Sun the total time spent on the main sequence is around ten billion years of which it has already spent around 4.6 billion years. The Sun has still a long way to go before it becomes a giant and threatens the Earth! For less massive stars the central temperatures do not rise high enough to initiate fusion of very heavy nuclei and so they do not produce all the elements up to the iron group. The case described in the above table relates to a massive star.

The origin of the elements

The above scenario of how a star functions as a nuclear reactor furnishes the answer to our fundamental question: Where do the chemical elements originate?

The stainless steel spoon you use for eating no doubt was made from metal mined from the Earth. But how did it get there? It must have been around within the primeval gas cloud from which the solar system condensed. But how did it get there? The answer proceeds as follows.

We now see that elements can be made within stars. The process described above as taking place within the red giants is called the α-process since the fusion of helium nuclei in succession leads to elements of increasingly heavy atoms — ^{12}C, ^{16}O, ^{20}Ne, ^{24}Mg and so on. The elements form what is called the alpha-ladder.

Elements not on the ladder can also be formed by different processes in the stars. The first comprehensive calculation of how nuclei are made in the stars was made in 1956 by Geoffrey and Margaret Burbidge, William Fowler and Fred Hoyle. These astrophysicists showed that not only the nuclei up to the iron group are made in the stars but also heavier ones; as well as those off the alpha ladder. We will come to these processes which are given names like e-process (equilibrium process), r-process (rapid process), s-process (slow process) etc. But before we come to that topic, we need to find some mechanism for the star to be able to deliver the various elements trapped deep in its interior. For this to happen, the star must break up somehow — that is, it should reach a state when it is no longer able to sustain its overall equilibrium. Will such an eventuality occur?

Yes, as we shall see in the next chapter, the star of large enough mass does reach a stage when it can no longer maintain equilibrium. It simply explodes. Indeed stellar explosions are necessary if the nuclei manufactured by stars deep inside their cores are ever to get out into the interstellar space where they form part of the material from which new stars and planets (like the Earth) are formed. Only then will our question "But how did it get there?" receive a complete answer.

Chapter 8

When Stars Explode

By applying the laws of physics discovered in the man-made laboratory we have been able to figure out how and why the stars shine, why they change their appearance from main sequence to the giant branch and how their internal composition changes. We also saw that their cores act as fusion reactors making the various chemical elements. But we have not yet reached the end of the star's life-story. For, we will find that some very dramatic events are in store for a star which has become a red giant. Before we describe them, let us look at a tale from three continents, which contains factual eyewitness accounts of the most catastrophic stage in a star's life. We begin with East Asia.

The tale of a "guest" star

The astronomers in China and Japan nine centuries ago were accustomed to keeping meticulous records of all heavenly bodies. Their purpose in doing so was largely astrological. For it was widely believed that if the ruler of the country strayed from the straight and narrow path of virtue and behaved badly, he would be warned of a possible punishment by God; and the warning would appear through some unusual happenings in the sky. So it was the duty of the court astronomer to keep a close watch on the heavens and to report anything uncommon which could be interpreted as a warning sign.

Against the well-regulated movements of stars and not so well regulated but still predictable movements of planets, the unusual occurrences usually consisted of eclipses of the Sun or the Moon, meteor showers or visits by comets. The astronomers during the Sung Dynasty were certainly not prepared for what they saw on July 4, 1054 A.D. (date as per today's calendar!). Their records reads[1]

on a Chi-Chou day in the fifth year of Chi-Ho reign period a "guest star" appeared at the south-east of Thien-Kaun measuring several inches. After more than a year it faded away.

Further details appeared in the records which went on for several weeks. The appearance of a new star would certainly cause comment, especially when it was bright enough to be visible by day being about five times as bright as Venus in the morning or evening! The star, however, did not stay bright for long and began to fade. In two years, it was not visible to the naked eye. Its location in the sky relative to the background of other stars was fixed, however. The Chinese and Japanese record of the position "several inches south-east of Thien Kaun" implies Zeta Tauri in the constellation of the Bull. Because the star made only a transient appearance in the heavens, it was called a "guest star".

Was this the sole instance of people reporting an unusual celestial phenomenon? Where else in the world did people see this remarkable event? It should have been seen in India, Europe and the Middle East and also in the American continents. Unfortunately, we do not have meticulous records available. European intellectuals being dominated by the Christian dogma that God created the Universe at one go in all its perfection may have found it difficult to accommodate this apparition in their theological framework and probably preferred to ignore or suppress it.

What about India, which was passing through its 'Golden era' of knowledge, especially in mathematics and astronomy? July is a month of heavy rainfall in India and the sky is often overcast. So one could argue that nothing may have been seen in the early stages. However, even in the Monsoon period, you do not get overcast skies all over the country. Also, according to the Chinese and Japanese records, the event

---

[1] "History of the Sung Dynasty" by Ho Peng Yoke (1962).

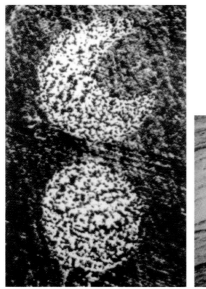

Figure 8.1: The pictures are from the rock records of Red Indian tribes in Navajo Canyon (left) and White Mesa (right). (Photograph by William C. Miller.)

lasted well past the Monsoon. A study organized by this author sponsored by the Indian National Science Academy, however, led to null finding after examining printed and handwritten manuscripts from the leading libraries all over India. One sociological explanation could be that recording new information was discouraged in the prevailing custom of learning by heart only the Vedic verses, passed on from teacher to pupil, in an oral tradition. The dictum was that true knowledge only came from the Vedas.

The Red Indians of America do seem to have left records of the event, records that were discovered by Bill Miller and discussed in print in 1955. The records are of two types, shown in Figure 8.1. The *pictograph* shown on the left is an image made on rock with paint or chalk (or a rock that is like chalk). The picture on the right is a *petroglyph* which is an image chiseled into the rock by a sharp object. The crescent in these pictures is, of course, the Moon. But what is the round object?

Although these pictures cannot be dated, there is good circumstantial evidence that the round object is the guest star. For, the Moon was crescent at the date of star's first reported sighting in China and Japan. Moreover, the star would have been close to the Moon as shown in the picture. Further, such carvings or pictures were found in locations from where the Eastern horizon was clearly visible or from where the view of the horizon was easily accessible. This circumstance acquires significance when we note that the above event was expected to be seen on the Eastern horizon.

In 1978, Kenneth Brecher, Elinor Lieber and Alfred Lieber uncovered evidence that the event in question was also seen and recorded in the Middle East. Ibn Buttan, a Christian physician from Baghdad who lived in Cairo until late 1052 or early 1053 and who later moved to Constantinople, records that a spectacular star was seen in Gemini sometime between 12 April, 1054 and 1 April, 1055. When adjustment is made for the change in direction of the Earth's polar axis over the last nine centuries this position in Gemini corresponds to the present position of the guest star in Taurus.[2]

Present position? Yes! If we look in the direction of where the ancients are supposed to have seen the spectacular guest star, we do not see anything today with our naked eyes; but photographs taken through a telescope do show something quite spectacular. Known as the Crab Nebula (see Figure 8.2) this object is believed to be the same as that which was seen in July 1054 in broad daylight; but with one difference! What we see today in this nebula is the *remnant* of the explosion seen on July 4, 1054 in which a star lost most of its outer envelope. Such exploding stars are called *supernovae*.

Supernovae

When does a star become supernova? This stage is reached in a star's life when it has gone to the limit of the nuclear fusion process, that is

---

[2]The North-South polar axis of the Earth is not fixed in space; it sweeps out a cone in a period of about 26,000 years. This movement of the axis is very much like the precession of the axis of a spinning top. As a result the overall backdrop of the night sky in relation to the Sun slowly changes with time.

Figure 8.2: The Crab Nebula in Taurus which is the remnant of a supernova explosion.

when its central core has manufactured the nuclei of the iron group. Indeed, that is the stage which we had brought our story to at the end of the last chapter, the star itself having taken the overall shape of a red giant. Something must happen now to disturb the star's equilibrium and cause it to explode.

With the manufacture of the iron group nuclei, the fusion process stops and the star's core begins to contract. This had happened to the star several times in the past: whenever it exhausted its fuels of hydrogen, helium, carbon etc., progressively. But on those occasions one fuel gave way to another and the process of fusion had carried on somehow. Now the situation is different. So far as nuclear fusion is concerned the fuel is fully exhausted.

So when the core contracts and heats up further, no nuclear fusion is triggered. Rather the reverse happens. The iron group of nuclei break up into alpha particles leading to a *loss* of energy in the core. This leads to the disruption of the envelope.

The process by which this happens is somewhat similar to the blowing of a balloon with an air pump: to fill up the balloon with air we pull back the piston of the pump and then push it in. Air trapped between the piston and the mouth of the balloon enters the latter at high pressure and blows the balloon up. Likewise, the core of the star first collapses under its force of gravity and then bounces back. And in so doing it exerts a tremendous pressure outwards on the envelope.

Why does the core bounce? We will return to this question in the next chapter when we consider the fate of the core. Let us assume for the time being that the core bounces because it encounters a hard resistance deep inside, which not only halts the infall of the outer portions of the core but also bounces it back.

The sudden reversal of the core velocity from inward to outward direction catches the envelope unawares. The result is described in technical jargon as a *shock wave*. What is a shock wave?

An ordinary sound wave travelling in air or water produces changes of pressure and temperature in the medium that are *small* and *continuous*. That is, the changes from any one point to a neighbouring point are smooth and they do not disrupt the overall pressure pattern of the surrounding medium. Not so in a shock wave! The shock wave arises whenever there are *sudden, large and discontinuous* changes of temperature and pressure between neighbouring points as can happen in an explosion. The abrupt change passes swiftly outwards through the star and produces two effects.

One effect is to suddenly heat up some of the outer portions of the star. Recalling the onion skin picture of Figure 7.7 we note that the outer portions have shells of less heavy elements like oxygen to silicon. These shells suddenly find themselves heated to temperatures of 4 billion degrees or so, as the shock wave passes through them. And, although the heating occurs for an extremely short time, say for not more than a few tenths of a second, it triggers off a burst of nuclear fusion. This fusion is explosive in character, although it does not produce many changes in the star's nuclear composition. Yet it can have some dramatic and lasting effects on the environment of the star, as we shall see later in this chapter.

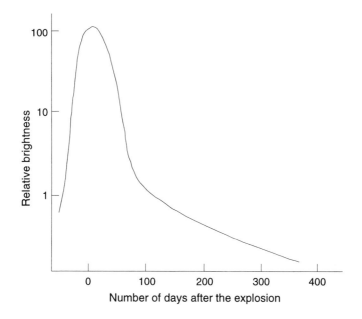

Figure 8.3: The curve shows schematically how the light intensity rapidly rises in the early stages of a supernova explosion and then declines slowly.

The second effect of the shock wave is to generate high outward velocities in the material comprising the star's envelope. The velocities generated are so large that the envelope (which was hitherto bound to the central core of the star by the mutual bond of gravity) breaks loose. This is the moment of truth for the star.

The sudden release and ejection of the envelope is the phenomenon known as the explosion of a supernova. In the initial stages of the explosion the star releases enormous energy in a very short time, so much so that in the brief moment of glory (before death!) it outshines the entire galaxy of hundreds of billions of steadily shining stars. No wonder then that such an explosion was visible on the Earth even during daytime on July 4, 1054 A.D.

Figure 8.3 shows how the light intensity rises sharply and then declines over a period of one to two years after the explosion. Like the Crab supernova of 1054, two other supernovae have been seen in our Galaxy, one by Tycho Brahe in 1572, and the other by Johannes Kepler in 1604. Many other supernovae have been seen in modern times but not in our Galaxy, even though there may be 2 to 3 supernovae exploding every century in it. The reason why not all these supernovae are seen is because their visible light gets strongly absorbed by the intervening interstellar dust (see Chapter 4).

The exploding stars not only emit great amounts of light, they also eject particles of matter at very high energies. The particles are electrons, neutrinos and the nuclei of atoms manufactured in the stellar interiors. The surrounding interstellar space therefore gets contaminated by these ejecta from the supernova. The heavy nuclei that are found in the universe, were thus processed in hot stellar cores and ejected in stellar explosions.[3] Thus the matter of which the Earth is made (and, of course, the constituents that we, humans, are made of!) indeed had a violent history.

Supernovae are believed to be sources of high energy particles that come in showers from outer space into the Earth's atmosphere. Known as *cosmic rays* these showers are detectable by instruments in the upper atmosphere to where they are carried by balloons, as well as by ground-based detectors. In addition to visible light and cosmic rays a supernova may also emit electromagnetic radiation of other wavelengths like radio and X-rays. Evidently, if a star within a distance of, say, thirty light years from us were to become a supernova, its ejecta would destroy life on Earth by eliminating the protective layers of its atmosphere like the ozone layer. (The Crab Nebula lies at the safe distance of 6000 light years!)

Looking at stars all over our Galaxy, we find that some stars contain very small quantities of metals. To clarify a possible confusion, all elements except hydrogen and helium are called 'metals' by astronomers! Stars with low metal content are called Population II stars. Stars with relatively high metal content are called Population I stars. These formed from the interstellar material which had already been enriched by the metals formed by Population II stars. The supernovae in latter population did this job. The Sun is a Population I star. Clearly, Population

---

[3]Stars of later generation formed from material containing supernova ejecta therefore have heavy nuclei right from the beginning. These help run the CNO cycle in the more massive stars described in Chapter 7.

II stars are older than Population I stars, as they formed earlier. But how did Population II stars get metal even in small quantities? It is believed that early in the universe there was an additional population of stars called Population III stars all of which were very massive. So they evolved fast and exploded within a million years or so, ejecting all the metals formed inside them. Out of this the Population II stars formed. So far no Population III stars have been seen but they are needed to understand the above issue.

Supernova 1987A

If not in our Galaxy, we do see supernovae go off in other nearby galaxies. Astronomers catalogue the supernovae so observed every year by an alphabetical sequence A, B, C, . . . , with the letter following the year in which the supernovae are observed; thus indicating their chronological order.

So the supernova catalogued under the name 1987A was the first supernova to be seen in 1987. Ian Shelton using a telescope in Chile photographed the eruption of the event on February 23, 1987, although, as we will shortly see, the news of the event had reached the Earth a few hours earlier by another means. This supernova turned out, in many ways, to be a fertile ground for testing the astrophysical theories of stellar interiors.

The star which exploded, Sanduleak (catalogue name Sk-69 202) was a blue supergiant star with a surface temperature of 20,000 K (on the absolute scale) and luminosity 40,000 times that of the Sun (see Figure 8.4). Its radius was estimated at 15 times the solar radius and its mass at the time of formation 17 times the solar mass.

These details could be estimated owing to the fortunate circumstance that this supernova was practically on the doorsteps of our Galaxy. It belonged to the Large Magellanic Cloud which is one of the small satellite galaxies going round the Milky Way. This galaxy derives its name from the historical circumstance that it was spotted by the world-traveller Ferdinand Magellan from Portugal while on one of his ocean voyages. The star itself was at a relatively modest distance of 50 kpc (= 170,000 light years) and was visible relatively easily.

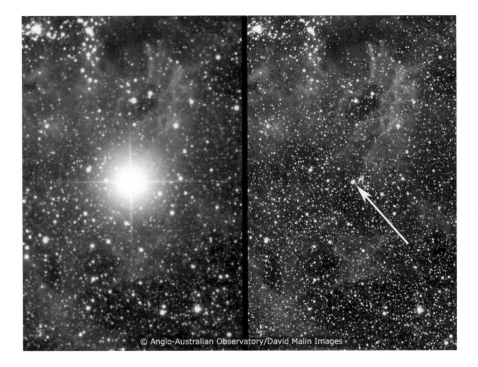

Figure 8.4: The star Sanduleak shown on the right exploded as a supernova shown on the left. This was the first of the supernovae seen in 1987 and is, therefore catalogued as Supernova 1987A.

Astrophysicists estimate that the collapse of the core which triggered off the explosion occurred a few hours before the explosion. If it were possible we should have witnessed the event at 07.35 Universal Time[4] on February 23, 1987. However, we cannot 'see' inside a star; but there is another way the information was in fact brought to us. *A large flux of neutrinos was released at the time.*

Neutrinos are subatomic particles which are believed to be 'massless' and like the photons, they travel with the speed of light. Unlike the photons, however, neutrinos have extremely weak interaction with matter. As a result, they can escape from the deep interior of an

---

[4]Universal Time is the clock used by astronomers worldwide to record events. It is the Greenwich Mean Time used earlier, with a few technical corrections.

about-to-explode star and travel outwards. (Recall that a photon produced in the deep interior of a star finds it very difficult to make its way
out and is scattered frequently. The neutrinos do not suffer that fate.)
Thus, before the actual explosion is to take place the sea of neutrinos
produced in the core of the star flows outwards in all directions.

As luck would have it, two laboratories, one in Kamiokande, Japan
and the other, known as IMB, in the United States had neutrino detectors already set up. Both detected 10 neutrinos each a few hours prior
to the visual sighting of the explosion. This was exactly as expected
theoretically.

The supernova 1987A was, of course, monitored optically by several
observers and its light emission increased rapidly in a day, to a thousand
times that of the original star. The radial size also increased from 15
solar radii to the size of the orbit of Mars. This was when it became a
supernova. When it was discovered by Sheldon optically, 22 hours had
elapsed since the core collapse.

The nuclei produced in a supernova include some which decay
through radioactivity. The decay products include high energy gamma
rays. Not all gamma rays escape without energy loss; but some do and
these were detected initially by the *Solar Max* satellite and later by
balloon experiments. In the last chapter we will return to the more
dramatic aspects of the so called 'gamma ray bursts' that have been
discovered in recent years.

Between the summer of 1987 and 1988 the total luminosity of the
supernova, arising from the gamma rays which lost energy to visible
light and infrared, declined exponentially with a mean life of 114 days.
These and other data provided valuable checks on theories of stellar
nucleosynthesis.

Thus the appearance of Supernova 1987A showed how in modern
times the astronomers, with many different checks can test and improve
their theories.

Planetary Nebulae

Not every star undergoes this traumatic fate. Just as on reaching middle
age, we humans are advised by our medical consultants not to exceed

Figure 8.5: The ring nebula shown here arose from the ejection of gaseous material from a star of low mass in a relatively minor explosion. The star is located at the centre of the ring.

specified weight limits in order to ensure a healthy life, so it seems to be the case for stars too! Stars reach their middle age when they become red giants and their subsequent stability depends on how massive they are. Calculations of stellar structure suggest that the very massive stars, say with more than six times the mass of the Sun, become supernovae. The less massive ones also undergo explosions but on a minor scale. These explosions eject material from the envelope in short puffs and the star gradually loses its mass. These puffs appear as rings lit up by the parent star and are called *planetary nebulae* in analogy with planets lit by their parent star. Figure 8.5 shows an example of a planetary nebula arising from such small scale explosions.

So far we have confined our attention to the more spectacular fate awaiting the outer portions of an exploding star. Does the star fragment entirely to pieces or does it leave some core portion behind? Although a very precise answer to this question is not known, it is generally believed that the tightly bound core portion of the star survives the catastrophe.

What are these left-over portions of stellar explosions? We defer the answer to the next chapter.

Before we leave the topic of exploding stars, let us look at some recent evidence suggesting that even these catastrophic events involving break up of a star do have a constructive role to play in the overall scheme of nature.

Star formation revisited

In Chapter 5 we discussed the current ideas on star formation and the origin of planetary systems. We are now able to answer the first question posed at the end of that chapter: that the formation of new stars from interstellar clouds could be assisted or even induced by the explosion of a nearby supernova. We will describe two lines of evidence that point to this idea.

The first piece of evidence was brought by a meteorite that fell in 1969 in the Mexican village of Pueblito de Allende. Known as the *Allende meteorite*, it showed certain peculiarities in its nuclear composition. It is these peculiarities, known as *isotopic anomalies*, that hold important clues to the origin of our Solar System.

Normally the nucleus of each chemical element is defined by the numbers of protons and neutrons it contains. Thus Carbon found in everyday life has six protons and six neutrons. Normally also, all chemical properties of the element are determined by its proton number. An *isotope* of an element contains a nucleus with the same number of protons but with a different number of neutrons. Thus, an isotope of carbon exists with eight neutrons and six protons. Its chemical properties are no different to those of the standard Carbon. In general the 'normal' or standard nucleus is stable while the isotopes are unstable and may break up sooner or later.

Our example in the present context relates to the metal aluminium which our pots and pans are made of. It is the stable element containing in its nucleus 13 protons and 14 neutrons. It is written as ^{27}Al. It has an unstable isotope called ^{26}Al which contains 13 protons and 13 neutrons in its nucleus. As mentioned before, the chemical properties of an element are determined by the number of charged particles in its nucleus, both ^{27}Al and ^{26}Al would have the same *chemical* properties. But their *nuclear* properties are different.

The unstable ^{26}Al is a radioactive substance and has a "half-life" of 720,000 years. That is, if we have in store 100 nuclei of ^{26}Al, on an average, half of them (50) would decay by radioactivity in this period. The main decay product is an isotope of another element called magnesium. The process of decay is written as follows:

$$^{26}\text{Al} \rightarrow {}^{26}\text{Mg} + e^+ + \nu.$$

The magnesium nucleus contains 12 protons and 14 neutrons. Thus one of the protons in the original nucleus is changed to a neutron. In addition, a positron (e^+) and a neutrino (ν) are released.

Now the Allende meteorite was found to contain certain isotopes in proportions very different from those normally found in the components of the Solar System. These differences of abundances are known as isotopic anomalies. Among these is the anomalously high proportion of ^{26}Mg. Why should this happen?

The above question and its answer can be better understood with an analogy. Suppose a country has imposed gold control laws under which its citizens are not permitted to hold pure gold beyond a prescribed quota. If a spot check of a section of the population turns up a person possessing gold way above this quota the question arises: how did that person acquire so much gold? Investigations may eventually lead to the discovery that he had smuggled the gold from another country where it was easily available. So the question astrophysicists asked about the Allende meteorite was: *Where and how did it acquire anomalously large stores of magnesium?* Their investigations which are described below were no less exciting than those unearthing the sources of contrabands.

There are many processes which could in principle produce the extra ^{26}Mg. However, a clue to the correct answer was revealed when the mineral grains of the meteorite were carefully analysed. It was then found that the abundance of ^{26}Mg was correlated with that of ^{27}Al, thus suggesting some link between magnesium and aluminium. As we have just seen, the link is via ^{26}Al which decays to ^{26}Mg.

So it was concluded that either the ^{26}Al somehow got into the meteorite material and then decayed there over the period of 720,000 years or so, or the meteorite was made from a portion of the interstellar medium already containing ^{26}Mg formed from the decay of the ^{26}Al present in the medium. The latter scenario seemed more feasible but it implied that the meteorite must have been formed *soon after* the contamination of the interstellar medium with ^{26}Mg; for otherwise the constant churning up of the medium by cosmic processes would have wiped out the signature of any *pristine* contamination. Hence the conclusion that the formation of the meteorite would have taken place soon after the deposit and decay of ^{26}Al in the interstellar medium. What cosmic process could have deposited this isotope of aluminium in the interstellar space?

This is where the supernova comes in. Notice first that the α-particle ladder we described in the last chapter increases the number of particles in the nucleus by steps of 4. Thus we get ^{12}C, ^{16}O, ^{20}Ne, ^{24}Mg and so on. ^{26}Al does not fit into this sequence. But it can be made during the explosive nucleosynthesis phase of the supernova, that we described earlier. In this phase free neutrons (n) and protons (p) can be added to form nuclei off the alpha particle ladder. Thus one possible way ^{26}Al is made from ^{24}Mg is by the series of reactions described below:

$$^{24}\text{Mg} + n \rightarrow {}^{25}\text{Mg}$$

$$^{25}\text{Mg} + n \rightarrow {}^{26}\text{Mg}$$

$$^{26}\text{Mg} + p \rightarrow {}^{26}\text{Al} + n.$$

There are also other ways of making ^{26}Al in this phase of the supernova. These ejecta from the explosion can very well contaminate the nearby interstellar space.

Therefore, the isotopic anomalies of the Allende meteorite, like that of ^{26}Mg which we discussed above, and others too suggest that they arose because a supernova went off in the neighbourhood of the gas cloud out of which the Solar System formed. And, the occurrence of the supernova cannot have been very far back in time before the formation of the Solar System. For example, if the time gap between the supernova explosion and the formation of the Solar System were say a million years or more, all signature of supernova contamination would have been wiped out.

This evidence from the Allende meteorite links the origin of our Solar System with a comparatively recently exploded supernova. It could still be possible that the presence of the supernova in the vicinity of the Solar System was purely accidental as also the timing of its explosion just before the Solar System began to form. However, supernovae being somewhat rare objects, there must be more to it than meets the eye. Indeed there is a physical argument to suggest that the explosion of a supernovae triggers off the process of star formation in its vicinity. Let us briefly look at this argument before examining the second piece of evidence.

Recall first that the explosion of the supernova was caused by a gigantic shock wave that originated in the star's core and travelled outwards. The wave in such situations does not terminate at the star's surface but continues moving outwards. As it recedes from the centre of explosion its intensity naturally declines. However, in the immediate vicinity of the supernova it may be quite powerful. Such a wave impinging on a nearby interstellar cloud can therefore give it a strong push. This push is just what is needed to set off the cloud's compression and it resolves the difficulty we mentioned at the end of Chapter 5: namely how to start the compression of a large diffuse cloud. The external pressure from the shock wave tilts the balance of all the forces on the gas in the cloud in favour of contraction. Do we have any evidence that such shock waves existed in the vicinity of newly forming stars? Yes! Just such evidence was uncovered in 1977 by two astronomers, W. Herbst and G.E. Assousa.

Herbst and Assousa examined the vicinity of the astronomical object called Canis Majoris R-1. As shown in Figure 8.6, it is a supernova-remnant like the Crab Nebula of Figure 8.2. As in the Crab, there is evidence of outward motion of gas particles, indicating that there was an explosion. Estimates show that the explosion took place about 800,000 years prior to the state now being observed in Canis Majoris R-1. More interestingly, new pre-main sequence stars have been seen in a region not too far from the supernova remnant. The stars whose ages are believed to be only about 300,000 years are probably the youngest stars known to astronomers.

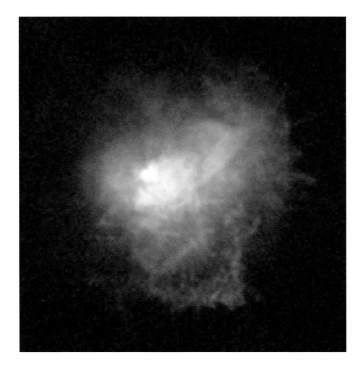

Figure 8.6: Photograph of the supernova remnant Canis Majoris R-1.

Evidently these stars formed *after* the explosion. How big was the explosion? If we try to work backwards from the present observations of outward moving gas, we arrive at a figure of 10^{44} joules for the energy released in the explosion. For a comparison, the Sun radiating with a power of 4×10^{26} watts will take around eight billion years to emit the above amount of energy. Fantastic though this figure seems in the normal stellar context it is characteristic of the energy in a supernova explosion.

Is there any evidence for the exploding star, say in the form of the leftover portion, that is, the inner core? We do see a star, not within the supernova remnant but outside it. This star is found to be moving away from the remnant with unusually high speed. Could this be the star whose envelope was ejected in the supernova explosion? A plausible case for such a course of event can be made on the analogy of a rifle shot.

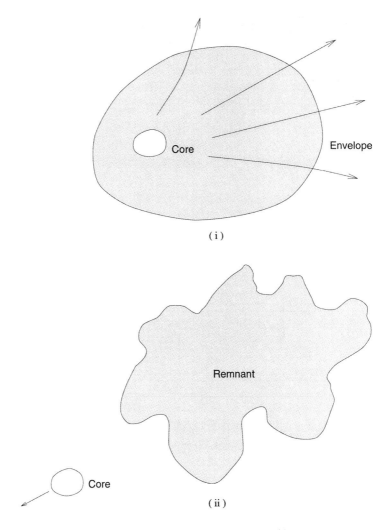

Figure 8.7: A skew explosion of a supernova shown in (i) can leave the remnant envelope well away from the surviving core which moves rapidly as in (ii) due to the recoil effect.

Just as the rifle recoils on firing the shot, the star in question recoiled after ejecting its envelope in the opposite direction. Figure 8.7 shows how a high recoil speed is generated in an oblong asymmetric explosion. The measured speed of the star does fit the recoil hypothesis.

So there is suggestive evidence that links the formation of new stars to a recent supernova explosion and it lends further strength to the hypothesis that star formation in GMC's is induced by the explosions of stars of an earlier generation. Our story of the star's life has thus come a full circle by linking the destruction of one star with the birth of another! One is reminded of the lines from T.S. Eliot: *In my beginning is my end, and in my end is my beginning.*

But there is more to come in the star's life even after its apparent destruction in the supernova explosion!

Chapter 9

Very Dense Stars

The sun has a mass of 2×10^{30} kg and its radius is 7×10^8 metres. If it were a homogeneous ball of matter its average density would be 1.4×10^3 kg per cubic metre, that is about 40% more than the density of water. Of course the Sun is not homogeneous and its density varies from the very low value of less than ten millionth part of the density of water at the outer surface (the photosphere) to about a hundred times the density of water at the centre.

There are, however, stars far more compact and far more dense than the Sun. Imagine the Sun to be compressed from all sides to a ball with only a hundredth of its present size. Its average density will rise a *million fold*. Such dense stars are the *white dwarfs* that we found in the lower left part of the H-R diagram. How do these stars form?

Going to even greater extremes, there are stars which are almost invisible in ordinary light and therefore do not find a place on the H-R diagram but whose radius is no more than a few tens of kilometres! If the Sun were compressed to a radius of 7 kilometres, that is by a factor of a *hundred thousand* from all directions, its density will rise to upwards of a *million billion* times the density of water. Known as *neutron stars*, these highly dense stars can be detected by other forms of radiation that they emit. Let us now examine how and when in a star's life such dense phases arise.

White dwarfs

In the last chapter we found that provided the star was not very massive to start with, say not more than six times the mass of the Sun, it would not have to undergo the traumatic experience of becoming a supernova. Rather, it would shed off the bulk of its mass in minor explosions (like those seen in the planetary nebulae) and then end up with a mass not too different from that of the Sun. At this stage because the leftover portion is the hot inner core of the original star, it is very hot at the surface but visually very faint, so that it moves to the lower left-hand corner of the H-R diagram: the part occupied by the *white dwarfs*. But how is the star now able to maintain its internal equilibrium?

At this stage we recall that till now the star was able to generate thermonuclear energy to keep its central part hot enough to provide large pressures: otherwise it would begin to contract under its own gravity. By the stage the nuclear reactors within the star are no longer functioning, how is it able to maintain its internal balance?

The answer to this question comes in an unexpected fashion. First we note that what has become a white dwarf was once the central core of a more massive star. The mean density of a white dwarf is therefore considerably higher than that in an ordinary star on the main sequence. Indeed, as we mentioned in the beginning, densities a million times that of water are not uncommon for a white dwarf so that one litre of its volume contains a thousand tons of matter!

When matter is so densely packed something unusual happens to it: it becomes *degenerate*.

To understand this adjective we have to go back once again to quantum theory, which, as we have seen describes the behaviour of matter at the microscopic level, at the level of molecules, atoms and atomic nuclei. When we consider the state of matter at very dense level, we necessarily intrude into the domain of this theory. Quantum considerations impose a restriction on how closely matter can be packed. Let us try to understand this restriction in the context of close packing of the electrons in the star: for it is that aspect that provides a new kind of internal pressure for the star's support.

Hitherto we have paid scant attention to the electrons in the star because we were mainly concerned with how the nuclei of its atoms combine and generate energy. But the atom has electrons as well as the nucleus. In fact, the atom is electrically neutral, so that it has as many (negatively charged) electrons in its outer part as it has (positively charged) protons in its nucleus. At high temperatures these electrons get detached from their nuclei (see Chapter 5). In a white dwarf star these free electrons also get compressed and of the various particles of matter they are the first to experience the effect of quantum theory referred to above.

This effect, known as "Pauli's exclusion principle" (so named after its discoverer, Wolfgang Pauli), states that in any given region no two electrons can exist in the same state. This rule, taken together with the quantum theoretical result that electrons are not "point" particles but are spread over a small volume, implies that we cannot pack together groups of electrons arbitrarily closely. But, first, what do we mean by 'state'?

The state of an electron is determined by its energy, momentum and intrinsic spin (see Figure 9.1). The energy of the electron is largely the energy of its motion while its momentum is the product of the mass and the velocity of the electron; essentially it tells us how fast and in what direction the electron is moving. The spin state tells us about the way the electron is spinning about its axis. Starting from the state of lowest energy up to any given higher level of energy the number of electron states available is limited by Pauli's principle. So if we start packing them from the lowest energy state we will discover that up to any given energy level only a specified number of electrons can be packed in and to put more electrons in we need to ascend the ladder of states of higher and higher energy.

Thus in a white dwarf star we have an assembly of electrons of varying energies starting from the lowest upwards. And a mixture of such electrons with different energies and momenta and spins generates a pressure of its own — a pressure that resists any further compression of the material. Under such a condition when all lower energy states are filled the electron assembly is said to have become degenerate (see Figure 9.2). Now, the lower the overall temperature of the electron gas,

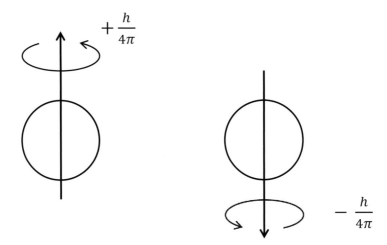

Figure 9.1: The electron has two discrete states of spin. When the angular momentum of the spinning electron in any specified direction is measured experimentally, we get two possible values $+h/4\pi$ and $-h/4\pi$, where h is Planck's constant.

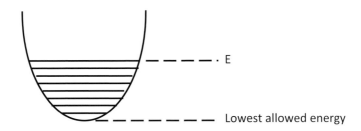

Figure 9.2: The above figure schematically illustrates electron degeneracy. The shaded region indicates that all energy states up to the level E are fully occupied. Any addition of new electrons is possible only at energies above E.

the sooner it becomes degenerate by compression. Thus the degeneracy pressure can come to the aid of a contracting star *provided* the electron gas in it is cool enough to become degenerate.

We can understand what is going on by looking at Figure 9.2 in analogy with the filling of a bucket with water. When the bucket is partially full of water, it is able to accommodate more. However, a stage comes when all of the bucket is full of water and no more can be accommodated. Likewise, where the matter is fully degenerate there is no more scope for extra matter to be pushed in the volume. In the bucket example, as more and more water is put in, the overall water level rises. So in the degenerate matter case, as all states of matter get filled up, the overall temperature rises in order to accommodate more electrons.

Let us now compare the fates of two stars, both of which are undergoing compression because they have no more nuclear fuel to provide sustained internal heat and pressure. Star B is massive while star A is light by comparison. The internal temperature in the star A is, as a rule, less than that of star B making it easier for the electron matter in star A to become degenerate than in star B. So we arrive at the qualitative conclusion that the less massive star is more likely to have degenerate gas pressure from electrons coming to its aid in halting contraction, than its more massive counterpart. Where is the dividing line between the "less" and "more" massive stars? We expect it to be a critical mass for such stars.

This dividing line below which the pressure of degenerate electrons can support a star in equilibrium was first calculated by S. Chandrasekhar in the early 1930's. He came to an answer that is now well known as the *Chandrasekhar limit* and it is approximately $1.44 M_\odot$. That is, stars with masses up to 44% more than the solar mass can keep themselves in balance and survive. Stars exceeding this limit cannot, and they will continue to contract further. Stars below the Chandrasekhar limit are of course the white dwarf stars. This was an important breakthrough in our understanding of very dense stars, and for this finding Chandrasekhar received a Nobel Prize in 1983.

A historic controversy

The critical argument in Chandrasekhar's reasoning was the difference in the physical behaviour of a degenerate gas when it is at low temperature and when it is at high temperature. In the latter case the electrons have high enough energies so that their speeds are not negligible compared to the speed of light. When this happens, it is necessary to take into account the effect of Einstein's special theory of relativity.[1] This is what Chandrasekhar did in his calculations.

Prior to Chandrasekhar's work, R.H. Fowler had also studied the behaviour of degenerate matter but in a non-relativistic way. According to Fowler's calculations, all stars, *whatever their mass*, would eventually generate sufficient internal pressure from degenerate electrons, to be able to settle down as white dwarfs. He had thus treated the electrons in very massive stars as moving as per Newton's laws of motion and not as per Einstein's special relativity. Chandrasekhar corrected this anomaly and his calculations using the effects of relativity altered Fowler's conclusion and brought into the picture the critical mass limit of $1.44M_\odot$. The notion of *relativistic degeneracy* was used for the first time in this calculation.

When Chandrasekhar presented this important result at the meeting of the Royal Astronomical Society, London on January 10, 1935, the work did not get the support and endorsement that it really deserved. The reason was that no less a person than Eddington expressed profound skepticism of the concept of "relativistic degeneracy" in Chandrasekhar's calculations.

Technical aspect apart, what disturbed Eddington most about Chandrasekhar's conclusion was the fate of a star that was unfortunate enough to exceed the Chandrasekhar mass limit and which could not therefore maintain its equilibrium against the contracting force of gravity. Commented Eddington: "The star has to go on radiating and radiating and contracting and contracting until, I suppose, it gets down to a few km radius, when gravity becomes strong enough to hold in the radiation, and the star can at last find peace... I think there should be a law of Nature to prevent a star from behaving in this absurd way!"

Eddington was right in visualizing the above fate of the 'unfortunate' massive stars; but he was wrong in presuming on behalf of "Nature". We will return to this comment in our next chapter.

---

[1]Special relativity introduces new concepts of how space and time measurements are made, concepts that differ significantly from the classical ideas of Newton and Galileo. The differences between classical and relativistic ideas can be ignored for particles moving *slowly* compared to light *but not otherwise*.

Although Eddington's ridicule was a setback to the Chandrasekhar limit graining immediate acceptance, it eventually came to be recognized as a correct piece of work. It is ironical that only a decade earlier Eddington himself was at the receiving end of criticism for his novel idea of nuclear energy as the source of starlight and had to wait several years to see it gain acceptance!

Why did the core bounce?

We now briefly revisit Chapter 8 and consider the cores left behind in a supernova explosion. These belonged to stars much more massive than those whose cores became white dwarfs. We have, therefore, to deal with states of matter hotter and denser than that found in a white dwarf.

To understand this state of matter let us take up the story of the massive star *before* it became a supernova. Recall from Chapter 8 the scenario that with the formation of the iron group of nuclei the fusion process ends and the core of the star begins to contract. At that time we stated that the collapsing core suddenly encounters resistance and bounces back. It was this bounce that caused the star to explode and shed its envelope.

What is it that causes the core to bounce?

The answer can now be given as follows. As the core contracts it begins to heat up. The supply of heat energy begins to break up the tightly bound nuclei of the iron group. This process is the reverse of the fusion process. Hitherto we had managed to extract energy by bringing lighter nuclei together to form the heavy nucleus. Now the heavy nucleus breaks apart by absorbing the energy provided by the heated core. The break-up of the nucleus releases free protons and neutrons.

Now a neutron in the laboratory cannot remain stable for a long time. If we have a group of free neutrons at any given time we will find that in a matter of around 12 minutes half of them will have decayed into protons, electrons and the antineutrinos. (The antineutrino and the positron are the respective antimatter counterparts of the neutrino and the electron.) The reaction may be written as

$$n \rightarrow p + e^- + \bar{\nu}.$$

(The minus sign suffix on e^- shows that the electron is negatively charged. The overhead bar of ν indicates that $\bar{\nu}$ is an antiparticle of ν.)

The free neutrons in the stellar core, however, do not decay. Rather, the reverse happens! The protons in the core combine with the loose (free) electrons to form more neutrons:

$$e^- + p \rightarrow n + \nu.$$

This process is called the *neutroniation* of matter. In principle a nuclear reaction like this can go also the reverse way. Local conditions (mainly the high density of protons) decide which way it goes. The build up of proton into neutron (rather than the decay of neutron to proton) does not happen normally in the terrestrial laboratory but is very common in the highly dense state of the matter in the contracting core. Thus very soon the core finds itself made up largely of neutrons.

These neutrons now play the same role in providing a degenerate pressure that the electrons did in the white dwarf. The same Pauli exclusion principle applies to neutrons which resist being tightly packed together. It is this resistance that is primarily responsible for the bouncing of the core (see Figure 9.3) prior to the supernova explosion.

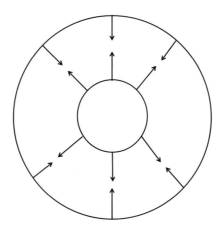

Figure 9.3: The hard centre of degenerate neutrons resists the collapse (shown by inward arrows) of the outer portion of the core and makes it bounce (shown by outward arrows).

Once the envelope has been shed off in the explosion, the core will recollapse and again the degenerate neutron pressure will be called into play. There may be another bounce and indeed the core may oscillate a few times before settling down to a static state, when there is exact balance between the degenerate pressure and gravity — provided, of course, that the total mass of the core is, again, not too large, above a critical limit.

For here we have a situation similar to that discovered by Chandrasekhar for white dwarfs. There is a limit to the mass of a star that can be supported by the degenerate neutrons. The limit is not very clearcut because the physical properties of matter at densities of million billion times the density of water were not well understood. But, now most workers in the field agree that the mass limit is close to $2M_\odot$. Stars with masses below this limit can maintain their equilibrium and are called *neutron stars*.

Figure 9.4 shows a schematic picture of how a neutron star is composed of different forms of matter ranging from a very dense state in the centre to a comparatively rarefied state in the outermost layers. It should be remembered, however, that even these rarefied outer layers are as dense as some of the inner layers of a white dwarf! Notice also that the star in Figure 9.4 is 40% more massive than the Sun but has an overall radius of only 16 kilometers.

How would we actually detect the neutron star? As we mentioned before, it is too faint and too hot at the surface to appear on the standard H-R diagram. Are there other ways of establishing its existence in a given part of the Galaxy?

In 1964, Fred Hoyle, John Wheeler and I had suggested in a paper in the scientific journal *Nature* that the neutron star might be detected through its oscillations. As mentioned earlier, the star is formed from the contracting core of a supernova and that the core oscillates before it settles down to a static form. Such oscillations of the star could continue for quite a long time since there is considerable dynamical energy to be got rid of. What we argued was that the energy could be dissipated by electromagnetic waves which are generated in the vicinity of the star through its oscillations. For, it is expected that a very large magnetic field would be present in the vicinity of the star which will take part in

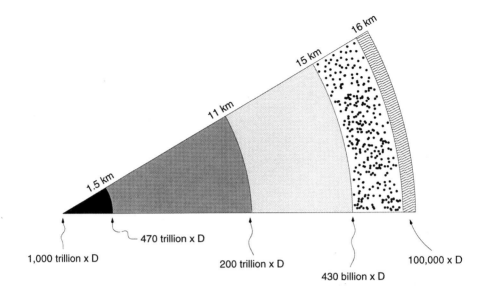

D ≡ Density of water

Figure 9.4: The wedge shows the varying composition of the neutron star based on a theoretical model. The figures below the horizontal line indicate how dense the above regions I-V are compared to water.

the osciliations and produce the electromagnetic waves. The wavelength of the radio waves as calculated by us was very long — about 3000 metres.

We had further argued that such long waves would be reflected back by any gas cloud with sufficiently high particle density. But in the process of reflection the waves would push the cloud along their original direction prior to reflection. So, according to this picture, the filaments in the Crab Nebula (Figure 8.2) appear to be moving away from the source.

As it turned out, many parts of the above scenario were correct. Thus the assumption of strong magnetic fields near neutron stars assumed in the above picture is now known to be true. A normal star may possess a weak magnetic field. But as it contracts, the magnetic lines of force passing through the star are squeezed as shown in Figure 9.5. Thus in the

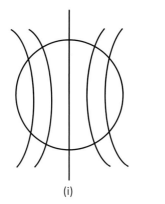

(i)   (ii)

Figure 9.5: As the star shrinks from (i) to (ii) the magnetic field (shown by lines of force: cf. Figure 5.10) gets compressed and intensified.

contracting core that becomes the neutron star, the squeezing is quite strong and this produces magnetic fields as high as a thousand billion gauss near the star's surface. (For comparison, the magnetic field near the Sun's surface is only 1 to 2 gauss. The Earth's magnetic field is even weaker.) It is also true that a neutron star is known to exist inside the Crab Nebula. But its presence was detected not through its oscillations but through its rotation, and that too in quite an unexpected fashion.

The discovery of pulsars

In 1967, Jocelyn Bell, a graduate student at the Mullard Radio-astronomy Observatory of the Cavendish Laboratory at Cambridge University, made a serendipitous discovery. While working on a problem of interplanetary scintillation, she noticed highly regular pulses of radiation arriving from a fixed direction in the sky. The pulses were approximately of 1.3 second duration.

It was very unusual for an astronomical source to send out pulses of such short duration. It was even more unusual for the period of the pulse to be so remarkably constant from pulse to pulse. Careful measurements revealed the pulse period to be

$$T = 1.3373011512 \text{ seconds.}$$

So many decimal places in the above figure testify to the high degree of precision with which the pulse period is maintained unchanged. Jocelyn Bell and her mentor Antony Hewish took pains to rule out both the down-to-earth explanation that the signals were from a terrestrial source being picked up by their very sensitive telescope, and the exotic alternative that these were the long awaited signals from advanced extraterrestrials! If the latter alternative were correct, then the ETs would be staying on a planet going round a star (for its energy source) and as such the frequency of the pulse should exhibit variable Doppler effect as the source orbited its star. No such Doppler variation of frequency was seen and it was concluded that the pulses were coming from a fixed source like a star. The signals were in fact identified as coming from a new type of astronomical source that was named as *pulsar*. Figure 9.6 shows the pulse pattern of the first Cambridge pulsar.

The shortness of the pulse period suggested that the source had to be small in size; yet it had to be powerful enough to be detected from so far away. With these clues theoreticians did not take long to home in on to the neutron star as the most likely candidate. A further support to the conjecture was provided a few months later by the discovery of a second pulsar in the Crab Nebula. In 1974 Hewish was awarded a Nobel Prize for the discovery of pulsars.

By now more than 600 pulsars are known, although only two (including the Crab pulsar) are clear-cut cases of pulsars found in a supernova remnant. It is likely that the outer shell of the supernova in most cases was ejected in an asymmetric fashion so that the core recoiled and moved away from the original site. We encountered an example of this type of ejection in Chapter 8. But what is the actual mechanism by which such short period pulses are being emitted by the pulsar?

What may be going on in and around the neutron star can be understood as per the following scenario first proposed by Tommy Gold in 1968. The neutron star has two polar axes: the rotational axis and the magnetic axis. The Earth also has two sets of poles, one from rotation and the other from the magnetic field. But unlike the case of the Earth where the two axes are nearly aligned, in a typical neutron star the two axes may be pointing in very different directions.

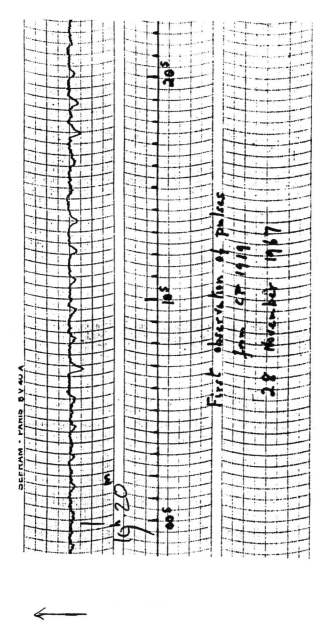

Figure 9.6: The discovery record of the first pulsar obtained by Hewish and Bell.

The rotating star has a swarm of electrically charged particles (the electrons) in its atmosphere. As the star rotates so does the atmosphere which is carried along by the star's strong gravitational pull. Just as the outer parts of a merry-go-round move much faster than its inner parts, the charged particles in the outer parts of the atmosphere move very fast, approaching the speed of light. Such fast particles are known to radiate electromagnetic waves in the presence of magnetic fields. Because of its similarity to radiation in man made high speed particle machines called *cyclotrons* the radiation itself is called cyclotron radiation. It is highly beamed like the beam of a rotating searchlight.

So if we happen to be located in the sweep-through area of the pulsar beam we will get pulses of radiation each time the beam sweeps past us, rather like the flashes of light periodically coming from the rotating beams of a light house. The pulse period is therefore just the period of rotation of the neutron star about its axis. In an alternative scenario first proposed by V. Radhakrishnan and D.J. Cooke the emission of the pulsar's radiation comes not from the star's outer atmosphere but from the surface of the star where its magnetic poles are situated. It is still too early to say that the details of a pulsar's emission process are fully understood.

Binary and Millisecond Pulsars

While the astronomers were getting used to the idea of stars spinning rapidly with a period of the order of a second, the 1980s saw the discovery of pulsars spinning considerably faster with periods measurable in milliseconds rather than seconds! It appears that in such cases the pulsar is not an isolated source but is part of a binary system. The fast spin develops in the pulsar as a result of accreting matter from its companion star which along with the pulsars forms the binary system. The binary and millisecond pulsars are comparatively rare, about a few percent of the total pulsar population. But they have generated considerable excitement amongst theoreticians for various reasons. One of them is described below.

A combination of a pulsar and another star going around each other as a binary system has proved to be an extremely valuable laboratory

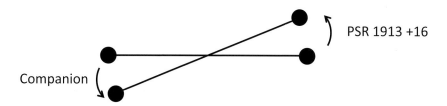

Figure 9.7: The schematic diagram shows that the line joining the binary pulsars PSR 1913 + 16 and the companion star when they are closest to one another gradually changes its direction in space.

for testing theories of gravity. For, as we saw, the pulsar provides a highly accurate time-keeping mechanism. If the changes in the binary system are studied with this natural clock, we can test the predictions of theories like Newton's law of gravitation, Einstein's general relativity and other competing theories of gravity.

The binary pulsar PSR 1913+16 discovered by Joe Taylor and R.A. Hulse in the 1970s provides such a laboratory. For example, Einstein's general relativity predicts that the orbit of relative motion of the two stars should not strictly be an ellipse as predicted by Newton's law. Instead, the orbit should steadily precess with the direction of closest separation between the stars slowly turning (see Figure 9.7). Such an effect was indeed observed for PSR 1913+16, thus confirming the relativistic prediction.

Another prediction of relativity is that the binary system should radiate energy in the form of gravitational waves. Just as the electromagnetic waves radiated by a light source carry away energy from the source, so do gravitational waves take energy away from the binary system. The result is a steady shrinkage of the relative orbit and a consequent decrease of period of the orbit. This decrease in the period is very small, but, thanks to the accurate time keeping by the pulsar, we can measure it. The result is in agreement with the predictions of the relativity theory. The importance of this work can be judged by the fact that Hulse and Taylor shared a Nobel Prize for physics in 1993 for their discovery of the binary pulsar.

The pulsar is probably the only example in astronomy where a star was detected first not by optical means but by using a different part of the overall spectrum of the electromagnetic waves. Shortage of space limits us from going into details of the many fascinating properties of pulsars but, in the *bibliography* at the end, we will refer the reader to sources where such information would be available.

Instead, we will now turn to find out what happens to those stars which are too massive to be supported as neutron stars and which have therefore neither the nuclear fuel nor the degeneracy pressure to keep them in balance. Do they follow the 'absurd' path described by Eddington?

Chapter 10

Black Holes

Right from its birth the star's life is a succession of struggles to maintain its internal balance gainst heavy odds. The odds of course are brought upon by the star's own weight — by the force of gravitational contraction generated by its own mass. Whether by nuclear fusion or by degenerate pressures the star does its best to counter this force, changing wherever necessary its internal composition and overall size. The different parts of the H-R diagram with the main sequence stars, the red giants, the white dwarfs and the invisible neutron stars indicate the extent to which the stars have to adapt themselves to changing circumstances. We saw in the preceding chapter that provided the final mass of the star does not exceed certain critical values it can continue to exist as a white dwarf or as a neutron star.

But what happens to those stars which are too massive to exist as white dwarfs or neutron stars during this final stage? Eddington (cf. Chapter 9) correctly forecast their future, although he felt that it was so bizzare that it may not actually be permitted by Nature.

The climate of opinion has changed since the 1930's when Eddington made this conjecture. This is not because Nature has revealed some dramatic secret, but because the theoretical astrophysicists of today have become bolder in proposing scenarios which would have been considered too radical for acceptance by an older generation. In so doing they have merely been following the dictum of Sherlock Holmes, the legendary detective created by Sir Arthur Conan Doyle: "When you have eliminated the impossible, whatever remains, *however improbable,* must be the truth."

So what Eddington thought an improbable scenario (that *a star would go on contracting until its gravity holds back all its radiation*) is probably one of the most popular astrophysical ideas of today. It goes under the name *black hole*. Before we come to discuss this extraordinary object, let us take a closer look at the force of nature that brings it about: the force of gravity. For, when a star becomes a black hole it finally concedes victory to gravity!

Gravity as a dictator

Although it was the first of the fundamental four interactions of Nature to be discovered and expressed as a mathematical law, gravity remains very much an enigma today. Newton's law of gravitation may be stated in the form

$$F = G\frac{m_1 m_2}{r^2} \tag{A}$$

This law states that the force of attraction F between two masses m_1 and m_2 separated by a distance r is expressed by the formula (A). G, a constant of Nature is called the "gravitational constant". When Newton was asked why such a law existed, his famous reply was *"Hypotheses non gingo"* (I do not make hypotheses). Newton's approach was an empirical one. The law of gravitation as given by formula (A) seemed to explain the motions of planets and satellites and could be tested further in other situations. The question as to why Nature produces such a law, requires a deeper reasoning which Newton refused to go into. Today, more than three centuries later the reasoning still eludes us.

We will, therefore, follow Newton, and take the law for what it is and look at some of its implications.

An immediate corollary of formula (A) is that the force of gravity is large if the two masses m_1 and m_2 are large. This explains why gravity as a force is almost negligible for atomic particles but is important in astronomy. Take for example, the law of electrostatic attraction between two electric charges $+e$ and $-e$ at a separation of r:

$$E = \frac{e^2}{r^2}. \tag{B}$$

Notice that both the electric force E and the gravitational force F are forces of attraction that increase in the same way as the particles approach each other. So the ratio of the two forces is the same at all values of their separation r:

$$\frac{E}{F} = \frac{e^2}{Gm_1m_2}.$$

For the electron-proton pair in the hydrogen atom this ratio is as large as *ten thousand billion billion billion billion* (10^{40}). This large number explains why atomic physicists do not bother to include gravitational effects into their calculation: it is far surpassed by the electromagnetic force.

But the astronomer deals with large masses which are electrically neutral. And so for him the electrical force $E = 0$ while the gravitational force F is very large. He cannot afford to leave gravity out of consideration. Indeed, we have already seen that throughout the star's life it is gravity that calls the tune.

An example will illustrate the dictatorial behaviour of gravity. Figure 10.1 shows two masses m_1 and m_2 connected by a rubber band. In Figure 10.1(a) the band is stretched so that m_1 and m_2 are far apart. But since the band tends to contract m_1 and m_2 feel a pull towards each other. Indeed, not for long! For, once they obey the dictates of the elastic force that tends to shorten the length of the band, the force itself diminishes and disappears altogether when the band attains its natural length. In Figure 10.1(b) we find that when m_1 and m_2 are closer to each other (and the band has attained its natural length) they experience no further force.

Figures 10.2(a) and (b) contrast the situation in the case of gravity. In (a) the masses are far apart and have a weak force of gravitational attraction between them. If, however, they give in to the dictates of this force and move closer together the force itself does not disappear. On the contrary it grows. In (b) therefore the masses m_1 and m_2, by coming closer, are experiencing an even stronger force than before.

Stretched spring

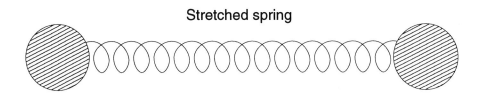

Normal spring

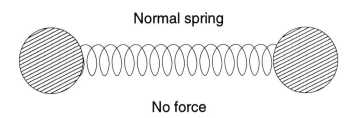

No force

Figure 10.1: The masses m_1 and m_2 are attracted towards each other in (a) because of the elastic tendency of the stretched band to contract. In (b) this force has disappeared when the band has shrunk to its natural length.

Gravitational force

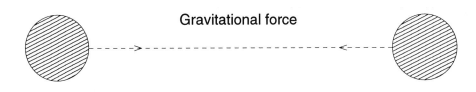

Stronger

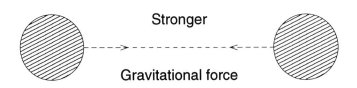

Gravitational force

Figure 10.2: The gravitational attraction between m_1 and m_2 increases as they come closer, in contrast to the situation depicted in Figure 10.1.

As Hermann Bondi has put it: gravity is like a dictator; give in to its demands and it will ask for even more.

This is what the star experiences throughout its life, especially so in the final stages. When the stellar core is contracting under gravity, it stimulates the behaviour of the two masses of Figure 10.2. As the core contracts, the overall force of gravity grows stronger and stronger and the star's hope of achieving structural balance recedes farther and farther away — for contraction means giving way to the demands of gravity, and giving way to the demands makes matters worse. This is why a star too massive to exist as a white dwarf or as a neutron star goes on contracting.

It should be emphasized here that this property of gravity makes it something of an exception in the context of forces in general. Forces in general behave like the elastic force shown in Figure 10.1; they efface themselves when their demands are met.

Let us now examine what happens to a contracting ball of matter which is not able to halt the process by any means whatsoever.

Gravitational collapse

Such a process of unstoppable contraction in which gravity gains increasing ascendancy is known as *gravitational collapse*. We had a brief encounter with this phenomenon in Chapter 6 when we considered the thought experiment in which the Sun was shrinking under its own gravity with no internal pressures to oppose the contraction. In that experiment we found that the Sun would shrink to a point in a period as short as twenty-nine minutes! This short time scale is an indication of how rapidly the force of gravity dictates the turn of events.

Let us follow that thought experiment a little further and ask ourselves the question: "Would we, as outside observers, be able to see this dramatic event all the way to its climax?" A little consideration will show that this will not be possible. Gravity, true to its dictatorial behaviour, imposes censorship on any signals going outwards to convey the information that the Sun has shrunk to a point.

Let us see why this happens. Consider first the example of the force of gravity we experience on the Earth. Figure 10.3 shows what

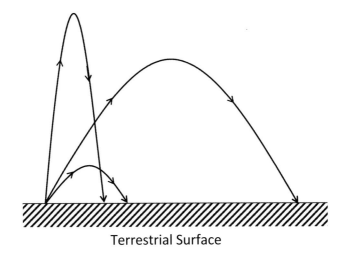

Terrestrial Surface

Figure 10.3: Objects thrown up in the air eventually come down to the surface of the Earth, pulled back by its gravity. Typical trajectories are shown above.

happens to any object that is thrown away from the surface of the Earth. Whether it is a pebble tossed up in the air or a cannon ball fired with great force, the object eventually comes down and goes back to the surface of the Earth, attracted by its gravity.

Does this mean that it is impossible to leave the surface of the Earth? That the answer to this question is 'no' is clear from the existence of powerful rockets like the one shown in Figure 10.4. These rockets are able to take spacecrafts beyond the confines of the Earth. Spacecrafts like the Pioneer 10 or the Voyager I & II have demonstrated that it is possible to leave the Earth with a speed that allows one never to come back.

Dynamical calculations show that it is possible for an object to leave the Earth forever if it is fired away with a certain minimum speed. This speed is called the *escape speed* and is close to 11.2 km per second (about 25,200 mph!). The largeness of the escape speed explains why the pebble and the cannon ball, which are fired up with much more modest speeds, fall back to the Earth. However, large though the escape speed is, it is attainable for man-made rockets, which explains why spacecrafts like the pioneer 10 could leave the Earth for ever.

Figure 10.4: The powerful Saturn rocket that launched the Apollo missions to the Moon.

The formula which determines the escape speed is rather simple. If we wish to know the escape speed V from the surface of any astronomical object of mass M and radius R, we use the formula

$$V = \sqrt{\frac{2GM}{R}}. \qquad \text{(C)}$$

For the Moon the relevant quantities in this formula are: for $M = 7.35 \times 10^{22}$ kg, $R = 1738$ km; and with $G = 6.66 \times 10^{-8}$ c.g.s. units, we get the escape speed as $V = 2.38$ km per sec. The escape speed from the Moon's surface is therefore much smaller than that from the Earth's surface. (This is fortunate; for the Apollo astronauts who landed on the Moon did not need large rocket power to enable them to leave the lunar surface for their journey back to the Earth.)

The formula (C) gives us a clue to the censorship property of a black hole. For, imagine what would happen to a massive body as it shrinks in size due to its own gravity. Its mass M stays constant but its radius

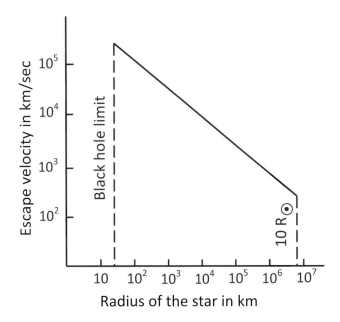

Figure 10.5: The escape speed of the ten solar mass increases to the speed of light c as the star shrinks from an initial radius of 7 million km to the black hole limit of about 30 km. (The plot above is on a logarithmic scale.)

R goes on decreasing. Figure 10.5 illustrates how the escape speed for a star ten times as massive as Sun increases as its radius R decreases. Notice that the escape speed is quite moderate for large values of R, say for $R = 10R_\odot$ ($R_\odot$ = Sun's radius = 700,000 km); but it rises rapidly and equals the speed of light

$$c \simeq 300,000 \text{ km per second}$$

at a radius of about 30 km. Thus as the object shrinks past this radius, not even light is allowed to escape from the surface of the body.

Since light is the swiftest (and sometimes the only) carrier of information from an astronomical body, it is clear from formula (C) that an external observer is denied any information about the object once it collapses inside the sphere of critical radius

$$R_S = \frac{2GM}{c^2}. \tag{D}$$

For an object ten times as massive as the Sun, this radius is about 30 km. We now see what Eddington meant about star's gravity becoming "strong enough to hold in the radiation", and why we won't be able to witness the Sun shrink to a point in our thought experiment.

Eddington was not right, however, about the peaceful future in store for the star: for the star's final moments are far from peaceful. To understand these final moments in the star's life we need to depart from the Newtonian theory of gravity and go to Einstein's general relativity.

Black holes according to general relativity

It will take us too far away from our main topic of stars to discuss in detail what general relativity is about. We will concentrate here only on those aspects of the theory which are relevant to the present situation.[1]

General relativity looks at the phenomenon of gravitation as a manifestation of geometry. By geometry, we mean, of course, the subject that deals with the measurements of lengths and angles of various figures drawn in space. The geometry that we learn in our school curriculum is that put together by the Greek mathematician Euclid about 24 centuries ago. Euclid's book starts with a set of assumptions (called axioms and postulates) and develops the entire structure of geometry through a series of theorems proved by rules of logic as *based on those assumptions.*

For a long time mathematicians thought Euclid's geometry was unique in the sense that there was no other geometry with different assumptions and different theorems. That this supposition was in error was realized in the 19th century and noted mathematicians like Lobatchevsky, Bolyai, Gauss and Riemann discovered new geometries which, as logical structures, were on par with Euclid's. So which geometry actually applies to measurements made in our space and time depends not on purely mathematical considerations. It is for the physicist to decide from actual measurements which geometry holds in nature: whether Euclidean or otherwise.

---

[1]For a detailed but nontechnical account see author's book, "The Lighter Side of Gravity" (Cambridge, 1995).

Einstein in 1915 proposed a physical theory with the supposition that the geometry of space and time containing concentrations of matter and energy is *non-Euclidean*. Known as *general relativity* this theory gives equations which relate the *non-Euclidean* aspects of geometry to the distribution of matter and energy in a quantitative way. We need not go into the quantitative details here but state a qualitative rule that the stronger the concentration of matter and energy in a given region the more radical is the departure of the rules of its geometry from those of Euclid's geometry. This approach can be illustrated by the following example.

Our textbooks in geometry prove a theorem that the three angles of a triangle add up to 180°, that is two right angles. But what about triangles drawn on the surface of the Earth? Imagine a traveller who begins the journey at the north pole [Point A in Figure 10.6] and walks

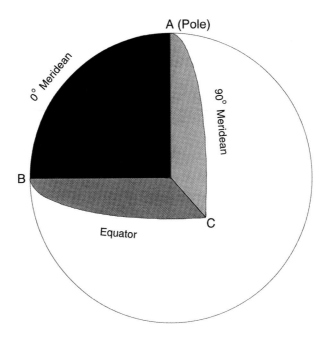

Figure 10.6: We have a triangle drawn on the surface of the Earth with the three angles adding upto 270°. The difference from the usual result arises because the geometry on the sphere is non-Euclidean.

straight in the southerly direction along the Greenwich meridian. This path will bring the traveller to point B on the equator. Here making a left turn the traveller proceeds to the East, one quarter way round the Earth, to point C which lies on the 90° meridian. At C the traveller turns northwards and going straight comes back to A. Now this trip is along three straight paths AB, BC and CA, but at each of the three vertices, the turning angle is 90°. So the three angles of our triangle add up to *three* right angles!

The reason why our terrestrial triangle appears to contradict the text book theorem is because that theorem is based on the axioms and postulates of Euclid's geometry all of which do not hold on the surface of the Earth. The postulate that is violated in this case is the so-called *parallel postulate*. In Euclid's geometry this postulate assumes that given a straight line l and a point P outside it one and only one straight line can be drawn through P parallel to l. We take it for granted as true. However, on the surface of the spherical Earth, any two 'straight lines' intersect. In short, *there are no parallel lines on the surface of the Earth*. Airlines flying over vast distances on the Earth require a different, non-Euclidean geometry, which applies on its curved surface.

Einstein argued that because of its gravity the space and time around a massive object become curved. Figure 10.7 shows a thought

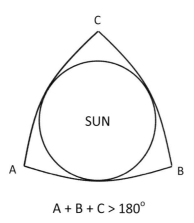

A + B + C > 180°

Figure 10.7: A triangle in space around the Sun should have its three angles adding upto slightly more than 180°, if general relativity is correct.

experiment illustrating this argument. Three space probes A, B, C around the Sun communicate with one another via light signals. Thus Figure 10.7 shows a triangle ABC drawn around the Sun with light signals. If we use sophisticated techniques of measurements we would find that the three angles of this triangle in space add up to slightly more than two right angles, implying that the space around the Sun is slightly curved.

The effects of this kind manifest themselves not only in spatial measurements but also in temporal ones. A typical effect is illustrated by another thought experiment. An observer B is located on a massive object where the gravitational effects are strong. Observer A is located far away from the object where its gravitational influence is weak. Then B's watch would appear to go slow compared to A's.

Suppose further that because of its strong gravitation, the massive object is shrinking and in this environment B agrees to send light signals to A every second as measured by B's own watch. A will not receive these signals at 1 second intervals but at intervals longer than 1 second. Moreover, these intervals will get longer and longer as the object shrinks; for with increasing density of matter around B, the geometry in the vicinity of the collapsing object becomes more and more non-Euclidean and so B's watch runs slower and slower compared to A's. As B approaches the barrier given by the radius $R = R_S$ in our formula (D), the gap between the successive signals received by A lengthens to infinity! In other words, the future of B as it crosses this barrier inwards will never be known to A even if A lives forever! This scenario is illustrated in Figure 10.8.

The significance of this barrier first became clear from a solution of Einstein's equations obtained by Karl Schwarzschild in 1916 and hence the barrier is often called the *Schwarzschild barrier* and its radius R_S the *Schwarzschild radius*.

Thus general relativity makes an even stronger statement (than the Newtonian law of gravitation) about the lack of communication with a black hole. Not only is the light prevented from reaching an outside observer, the apparent slowing down of time on the surface of the object implies that the observer in fact never lives to see the object become a black hole!

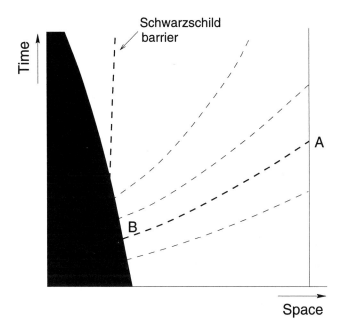

Figure 10.8: The above spacetime diagram shows the different rates at which the clocks of observers A and B run at different times. The horizontal displacement indicates distance from the centre of the collapsing object while the vertical axis measures time intervals. The shaded region is the shrinking boundary of the collapsing object on which B is located. Light signals at equal intervals from B shown by broken lines reach A at increasing time gaps by A's watch. The signals sent by B at or inside the Schwarzschild barrier never reach A.

To conclude this relativistic discussion, let us see what happens to the observer B after it crosses into the Schwarzschild barrier. Powerful theorems in general relativity derived in the 1960's by Roger Penrose, Stephen Hawking and Robert Geroch tell us that no conventional physical force can now halt the collapse of the object to a point. And at a point-like state the density of matter becomes infinite. The non-Euclidean nature of geometry near the collapsing object becomes stranger and stranger to the extent that in the end it defies any mathematical description! This final state is rightly called a *singular* state, and is supposed to mark the end of any physical description.

This singular end state is naturally hidden from the external observer A by the Schwarzschild barrier. But to the observer B the end is an unavoidable reality. Further, according to B's watch that end is reached soon after crossing the Schwarzschild barrier. How long does the entire collapse to a point take as measured by B's watch? The answer, somewhat surprisingly, is the same as what estimated earlier from the Newtonian theory: twenty-nine minutes for the Sun to shrink to a point if it were deprived of all pressures. In other words, though A may be blissfully unaware what goes on in the black hole, to B the ultimate end is a stark and imminent reality to be faced in a comparatively short time.

It is believed that such singular ends to collapsing objects are always shrouded in barriers like the Schwarzschild barrier in the above example. This belief is known as the hypothesis of *cosmic censorship*. Recently, however, some theoretical models have emerged which violate the cosmic censorship hypothesis. Such models allow an external observer to 'see' the singularity when it is formed.

The search for black holes

An object that, by definition cannot be seen, is naturally hard to locate. How do astronomers go about looking for black holes? Clearly, for the black hole to be taken seriously as a physical entity, its existence must be proved by astronomical observations.

It is true that a black hole cannot be seen by any of the telescopes available to the astronomers, ranging from radio telescopes, all the way to the gamma ray detectors. Nevertheless indirect methods can be used, methods that depend on the *gravitational* effects of a black hole on the surrounding matter.

An ideal situation in this respect is provided by double-stars. Figure 10.9 shows a binary star-pair, A, B, which consists of stars going round each other. In such a situation, the observer sees the positions of A and B change in space with a regular period. Their centre of mass C, however stays fixed (or moves with uniform velocity) at all times.

Suppose now A and B are fairly close to each other in the sense that the separation between A and B is not much greater than the

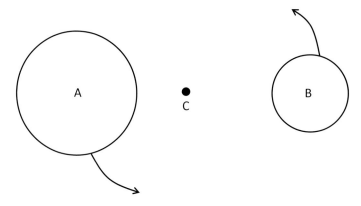

Figure 10.9: A pair of stars A and B going round their centre of mass C constitutes a *binary* star system.

sum of their radii. With stars so close to one another, each one tries to gravitationally pull out matter from the surface of its neighbour.

Such an interaction is known as *tidal interaction*. For, it is similar to the force that the Moon exerts on the Earth, causing tides in the oceans of the Earth. Thus when star B exerts a tidal force on A, matter from A nearest to B begins to flow towards B, and vice versa.

Now imagine a situation in which A is a giant star and B is a black hole. Provided A is near enough to B, matter will flow from A to B *but not vice versa*. For, matter from the black hole is never allowed to escape from it. This scenario is described in Figure 10.10. Here matter pulled out from A does not fall directly into B but goes round and round it before finally dropping in. This happens because both the stars A and B are going round each other and hence any matter leaving A has a tendency to go *round* B rather than fall directly onto it. Before we proceed further with this picture, one important observational aspect needs to be pointed out.

Imagine what this binary combination would look like if one of its members were a black hole. In that case we would see only one member (A in this example) going round and round. Seeing an isolated star going round without its companion would be inexplicable because as per Newton's first law of motion such a star should have gone on in a straight line with uniform speed. So the apparent isolation of a star

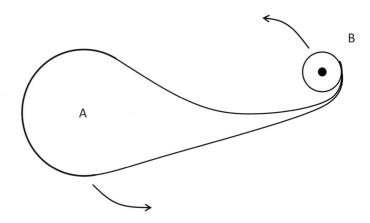

Figure 10.10: In a binary X-ray source the X-radiation comes from an accretion disc around the compact star B (shown above by the dark spot). The disc is formed by matter attracted by B from the surface of its companion star A. The arrows indicate the rotation of the binary system.

going round in a circle should excite our suspicion that there may be a companion present which we cannot see.

This infalling swirl of matter forms a disc-like structure which may extend from the black hole up to a distance several times the Schwarzschild radius. Since this infalling matter is in the form of gas or plasma, which is very dense and hot (because of frequent collisions of its atoms with one another), it radiates, largely in the X-rays. Several astrophysicists in the 1960's worked on this notion of an *accretion disc* round a black hole in the binary system: for, the newly emerging science of X-ray astronomy extended the hope of detecting black holes this way.

There is, however, one snag in this approach. What we have said so far about black holes also holds for a neutron star. Thus, if B in Figure 10.10 were a neutron star, it would also acquire an accretion disc around it, a disc that would radiate X-rays. And a neutron star would be invisible to optical observations.

So if we did find an X-ray source, associated with a binary system, only one member of which is visible then we could conclude that the other member is either a neutron star or a black hole. But how do we know which of the two we are concerned with?

Table 10.1: Binary X-ray sources in the Galaxy*.

X-ray Source	Mass limits on the compact component in units of $M_\odot$
3U 0900-40	1.6-2.4
Cen X-3	0.7-4.4
Her X-1	0.4-2.2
3U 1700-37	0.6-
Cyg X-1	6-15

* Partial list for which mass limits are reasonably well known.

It is here that we bring in the mass limit $2M_\odot$ for stable neutron stars. If by observations of the motion of the visible component A we are able to determine the mass of its companion B, and if that mass turns out to be below $2M_\odot$ then we can assert that B is a neutron star. On the other hand if the mass of B turns out to be considerably in excess of $2M_\odot$, we have reason to suspect that we are dealing with a black hole. In practice, the mass of B cannot be determined exactly but the observations of the parameters of A's orbit enable us to put a lower limit on the mass of B.

An additional check may come by monitoring the fluctuations in the X-ray radiation from the binary source. The accretion disc that is supposed to emit X-rays, does not maintain a steady shape and size. It is subject to fluctuations. More rapid fluctuations indicate a smaller accretion disc. Since black holes can be more compact than neutron stars, their accretion discs are also relatively smaller. Thus we expect very rapid variations of X-rays to arise from a black hole.

Table 10.1 below gives the data on some of the X-ray binaries in the Galaxy. Notice that the mass estimates of B in most cases are consistent with its being a neutron star. Only in two cases of this table (the first and the last) do we find some evidence for B being a black hole. Of these the most discussed case is Cygnus X-1. It is a binary of period 5.6 days, with A a large star of mass at least $9M_\odot$ while B has a mass at least *six times* the mass of the Sun. In addition, the very short time scale (of about the thousandth part of a second) in the variation of X-ray

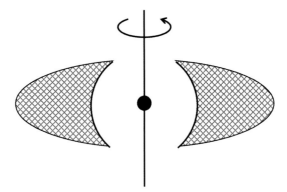

Figure 10.11: The cross hatched portions denote the axial section of a thick accretion disc formed around the central rotating black hole.

luminosity suggests that the disc is very compact and hence more likely to be around a black hole than around a neutron star.

Thus, it looks as if Cygnus X-1 is the only good clear cut case we have for a black hole. As any circumstantial evidence, this one also is subject to some doubts and a few skeptics would argue that the case for a black hole is not yet made. But then at present no other plausible interpretation of Cygnus X-1, *not involving a black hole* has been offered either. Therefore, the supporters of the black hole interpretation can justifiably claim their theory, being the only plausible one, must be taken seriously.

Apart from the binary systems, black holes have been invoked by theoreticians in many different contexts. Since the black hole represents a dense concentration of matter exercising a strong gravitational pull on the surrounding medium, it can in principle be a strong source of energy. For example, if a super-massive black hole of hundred million solar masses is spinning about an axis, it would pull surrounding matter into a thick accretion disc (see Figure 10.11) which can radiate powerfully. It is believed by many theoreticians that such a source must be responsible for the radiation from *quasars* — objects situated outside our Galaxy, which resemble stars but are in fact much more powerful radiators of energy than stars. A typical quasar may emit energy at a rate 10^{11} times that emitted by the Sun.

The example of quasars illustrates the importance of gravity as an energy reservoir. While stars mostly rely on their nuclear reactors to generate their luminosities, the same trick does not work for quasars. Quasars are much more luminous than stars but are comparatively much more compact. Their energy machines have to be very compact, powerful and efficient. Black holes of the supermassive kind appear to fill the bill.

Chapter 11

Binary and Variable Stars

We have nearly reached the end of our stellar odyssey. By adapting Purna's technique for the study of humans in Chapter 1, we have managed to build a reasonably coherent picture of how stars form, why and how long they shine, why they change their outward appearance, and how they reach the end of their lives. Nuclear physics and gravity hold the key to our understanding of stars. We have seen how stars with different masses end up differently, as white dwarfs, neutron stars, or the more exotic black holes. As a bonus for our studies of stellar evolution we got the answer to another fundamental question: How are the chemical elements made in the universe? There are, however, important gaps in the story still to be filled up. We will look at some in this chapter.

To begin with, we will introduce the subject of binary stars. When we look at the star-studded night sky we see hundreds of stars, each one appearing as a shining and twinkling point source of light. More careful observations, especially using telescopes will reveal not all of them are single stars: rather quite a few of them are star-pairs. Further observations will show that they are going round each other. Indeed, as per Newton's laws of motion any two gravitating masses near each other will go round each other. Generally called *binary stars* their studies have added considerably to our overall understanding of stellar populations.

Amongst the early observations of star pairs, we may mention the stars Mizar and Acrux. Situated in the Ursa Major (the big dipper) constellation, Mizar was observed as having a companion by Giovanni Battista Riccioli in 1650. Acrux is located in the Southern Cross and was seen as a binary by Father Fontenay in 1685. In general we may say that the invention of the telescope in 1609 brought about a phase transition in astronomical observations and binary stars are one consequence of the telescope revolution.

Binary stars

How common are binary stars? We may say that roughly half of all stars seen in the sky are members of binary systems; which makes the phenomenon quite a frequent one. The binary stars are sometimes referred to as 'double stars'. However, double stars may include pairs of stars which are seen to be very close to each other, but that may not signify that they are gravitationally bound systems. Rather one star while passing by happened to be close to another at the time of observation and it would go away in due course of time. An analogy with humans is that of a guest who stays with a family for a few days but is not part of that family. In this chapter we will not discuss such cases.

The binary population itself is classified into several categories, however. These are briefly listed below.

Visual Binaries: As the name implies this type of binary system is seen through a telescope or even through binoculars as two separate images of the member stars. Of course, a powerful telescope can pick up these separate images more clearly than a weaker one. As telescopic observations improve, more and more images get 'resolved' as pairs, thus increasing the number of visual binaries. In general the brighter the stars the more difficult are they to resolve as separate images.

The brighter member of a binary pair is called the *primary* star and the fainter companion is called the *secondary*. The stars are often labelled A and B respectively. Thus the binary pair Sirius is made of stars Sirius A and Sirius B. The former is a bright main sequence star while the latter is a white dwarf.

The stars are expected to move round each other in orbits following Kepler's laws of planetary motion. When interpreted in terms of

Newtonian dynamics, the barycentre (i.e., the centre of mass) of the pair stays in one place or moves with uniform velocity. Observations of the positions of the two stars should show their Keplerian elliptical orbits projected on the sky. The relative separation of the stars would likewise describe an *apparent ellipse* projected on the sky. Such observations enable us to determine the basic dynamical information of the binary system, such as the mass ratios of the stars, variation of their angular separation and orbital period. More information is available, of course, if we know the distance of the binary stars.

Spectroscopic Binaries: Often the two members of a binary system are so close that they cannot be resolved into separate stars even with the best of the telescopes at hand. In that case spectroscopy comes to our help. We can take the spectrum of the binary and it will contain a superposition of lines from the spectra of both the stars. However, at any given time there will be one star going away and another approaching us. So by Doppler effect, one set of lines will shift to the longer wavelength end, the *red end* and the other to the *blue or violet end*. As the stars move around these lines will oscillate between the two ends. This will tell us that we are looking at a binary system. Such a system is called a spectroscopic binary. Of course, the key source of information, the Doppler effect can identify only the projected radial velocity component of each star.

Evidently, spectroscopic binaries are easier to spot if the stars are moving fast, which they generally do in compact systems. If spectral lines from both stars are visible, such a system is known as a *double-lined binary*, often denoted by SB2. In case, because of faintness of the other star, lines from only one star are seen, the system is called a *single-lined binary* and denoted by SB1.

Eclipsing binaries: Suppose we as observers happen to lie in the orbital plane of a binary system. In that case we may be able to see a partial or total eclipse of one star by the other. These eclipses will alternately relate to the two stars in the binary. Such a case represents an example of eclipsing binaries. Because of the eclipses, the light from the eclipsed star goes up and down and one gets the impression of looking at a variable star. A good example of an eclipsing binary is the star Algol.

Astrometric binaries: The X-ray binaries described in the last chapter showed how one can *infer* the presence of an invisible member of a binary system by observing the orbital motion of the visible companion. In general such a binary system is called an *astrometric binary*. As we shall see later, this technique is useful also in the campaign for looking for extra-solar planets.

Of special interest to those trying to detect gravitational radiation, are the binary systems in which the two stars are very close to each other, almost on the point of coalescing. This is because typically a binary star is a steady source of gravitational waves and the intensity of the source rises if it has a high third time derivative of its quadrupole moment. [Compare the situation with that in electromagnetism where the basic source of electromagnetic waves is the second time derivative of the electric dipole moment of the source.] Coalescing binaries are thus ideal sources for gravitational radiation.

Binary evolution

It is likely that when star formation takes place, not one or two but a large number of new stars are born. These move in a cluster and during their mass movement, there are occasions when two stars come close enough to be trapped into a double system by their mutual gravitational attraction. It is also found from numerical simulations, for example, if *three* such stars get together, their mutual interaction will eventually provide enough kinetic energy to one of them to enable it to 'run away' leaving the other two more tightly bound into a binary system.

However, binary stars have presented interesting scenarios to workers in stellar evolution. At first it might appear that the two stars in a binary system evolve independently, each according to its mass, just as we have discussed in the preceding chapters. This expectation is correct provided the stars are well apart from each other. However, for close binary systems the situation is different! An important new development in such a case is of *mass exchange*. We illustrate the case with a specific example.

Figure 11.1 illustrates the situation. We have two stars A and B with masses $8M_\odot$ and $20M_\odot$ respectively. As B is much more massive

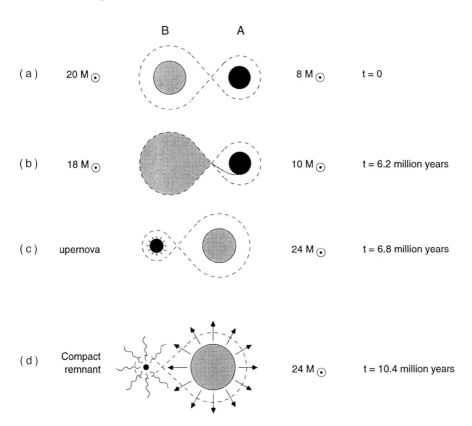

Figure 11.1: Two stars of masses $8M_\odot$ and $20M_\odot$ go around each other in a close binary system. As described in the text, mass exchange modifies the evolution of each star.

it evolves faster and takes about 6.2 million years to become a giant star. At this stage its size grows. Notice the dotted figure of eight that surrounds the two stars. It is an equipotential surface for the gravitational field of the rotating system. As the star B grows in size, this surface becomes important. It is called the *Roche lobe* of the binary and it determines the stage when the tidal interaction between the two stars becomes significant. It is named after the physicist E. Roche who first pointed out this result. So as B grows, matter begins to flow out of the Roche lobe towards A whose gravity pulls it towards itself.

After some time, around 6.8 million years from the start, B becomes a supernova and it explodes, leaving behind a neutron star core of mass around $2M_\odot$. What has happened to star A in the meantime? It has grown in mass to some 24 $M_\odot$ because of matter captured from its companion. This accelerates its evolution to the giant stage. In fact it is called a *supergiant* because of its large mass and matter from its surface begins to cross towards B in the form of what may be called 'stellar wind' which then leads to emission of X-rays through the formation of an accretion disc around B in the manner described in the previous chapter.

When the binary system contains a compact object like a white dwarf, or a neutron star or a black hole, gas from the companion accretes on to the compact member. Where the compact object is a white dwarf, the gas (which has become hot by friction or viscosity) emits optical radiation. In such cases the star is identified as a *cataclysmic variable*. The variability arises because of the instability of the accretion process and in some cases the sudden infall of the accretion disc on the star may ignite nuclear fusion reactions leading to cataclysmic bursts. In the cases where the compact object is a neutron star or a black hole the emission from accreted matter is largely in the form of X-rays as we saw towards the end of the last chapter.

Variable stars

Normally a star, like our Sun, keeps a steady profile as it shines in the sky. However, a few stars, because of internal or external reasons will show a variability in its brightness or in size and other physical characteristics. These are called *Variable stars* and they are further classified as *intrinsic* or *extrinsic* variables. The former refer to those stars whose luminosity actually changes because of internal physical changes like changes of pressure, rate of nuclear reactions, etc. The extrinsic variables may not themselves change but their physical environment may change, thereby varying the amount of starlight reaching the observer. We saw an example of this type in eclipsing binaries: here the eclipsing companion controls how much of the star's light reaches the observer.

Variable stars are named and catalogued just like ordinary stars. The convention followed in their case is as follows. The German astronomer, Bayer, had evolved a system of naming ordinary stars in a constellation by a letter in Greek alphabet followed by the name of the constellation in Greek, with the letter further down the alphabet used for progressively fainter star. Thus α *Draconis* is the brightest star in Draconis. After he ran out of the Greek letters he used the Latin ones and reached as far as Q. The first letter in alphabet not used in naming standard stars by Bayer was R. So the system developed for double stars by Argelander was to start with that letter for new variable stars. Subsequent discoveries in that constellation were named with prefixes RR, RS, ..., etc. up to RZ; then SS, ST, ...SZ, and so on. Thus RR Lyrae would be the first variable star in the Lyrae. Later discoveries used AA, ...AZ, BB, ...BZ, ... all the way to QZ. (For some reason J was omitted!). This took care of up to 334 stars. Beyond this number, the stars were simply labelled V335, V336, ..., etc.

Intrinsic variables

A very important group of variable stars are distinguished by their (internal) property of pulsation. A pulsating variable oscillates between minimum and maximum radii. This pulsation is accompanied by pulsation of luminosity and also of the spectrum of the star. This group of stars can be further subdivided into two classes: short and long period variables.

Cepheid Variables: These stars have a short period of pulsation, usually between days to months. The physics of pulsation is related to the concept of the so-called *Eddington valve*. The star has a layer of helium with varying level of ionization. When ionization is at a high level, there are more free electrons to scatter any passing radiation and so the layer has a high level of opacity. If the ionization is low, opacity drops. In the pulsating star, at low radius ionization is high and so the resulting high opacity means the layer absorbs more of the fusion energy. This increases the pressure and makes the star expand. As its density drops its ionization is also lowered and so a stage comes when the fusion radiation freely passes through the helium layer leading to lowering of the

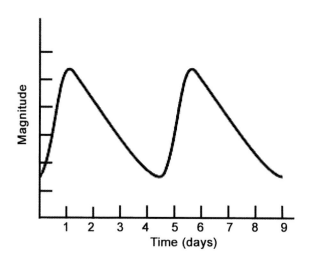

Figure 11.2: The Cepheid variable star shows periodic changes in its size, colour and light emitted per second. It is seen to be pulsating by measurements of the speed of its surface coming towards the observer and going away from them. These changes are indicated against time. The changes typically take place over a period of a few days. Here the graph shows how the luminosity of a Cepheid pulsates.

pressure. This in turn makes the star shrink under its gravity. Thus we have a feedback in both the expanding and the contracting stage. This is known as the Eddington pulsation mechanism and it drives pulsating stars like the Cepheids, RR Lyrae variables and the Delta Scuti stars.

The early discoveries of Cepheid prototypes include the discovery of Delta Cephei by John Goodricke in 1784. Although another similar star (Eta Aquilae) had been discovered (by Edward Pigott) a few months earlier, the genre became named after the second object to be discovered. The cepheids were studied in great detail by one of the pioneer woman astronomers in Harvard, Henrietta Swan Leavitt who discovered a relationship between the pulsation period and luminosity of these stars. Basically it tells us that the longer the period the more luminous the star. Figure 11.2 shows how a typical Cepheid changes its physical characteristics during a cycle of pulsation. The period-luminosity relationship is illustrated by Figure 11.3.

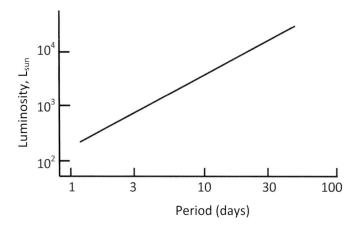

Figure 11.3: The changes taking place in a Cepheid variable have a fixed period, measurable in days. The longer the period, the more powerful is the star in terms of the light it emits. This figure shows how the absolute brightness increases with period, a relation found by Henrietta Leavitt in 1912. The luminosity is shown in relation to that of the Sun.

This relationship is extremely useful in estimating distances of galaxies. Because of its pulsation a cepheid in a far away galaxy can be spotted relatively easily. We can plot its apparent magnitude as it varies during the cycle. Since its absolute magnitude is determined from the period by the period-luminosity relation, we know its distance d in parsecs from the formula

$$M = m + 5 - 5 \log d.$$

There is another family of Cepheids called Type II Cepheids with a period luminosity relation different from the classical Cepheids just mentioned. Initially they were confused for the classical Cepheids thus resulting in wrong distance estimates. Originally they were named after the prototype star W Virginis.

Another intrinsic variable star is RR Lyrae. It is typically not as luminous as the cepheid, although it tends to be older. They also have a period-luminosity relationship and so can be used as distance indicators. These stars are spectral type A and they vary in luminosity by 0.2 to 2.0 magnitude over a period of a few hours to around a day.

There are more types of variable stars but we will not have space here to describe them all. We mention a few by names: RV Tauri, Pulsating white dwarfs, Cataclysmic variables, Mira variables, Delta Scuti, Dwarf novae, etc. . . .

Extrinsic variables

Amongst stars showing variability because of external (environmental) changes some important types are mentioned below.

Rotating variables: Suppose a star rotates about an axis. If it has spheroidal shape the area exposed to a random observer will change as the star spins and consequently the light received from it will also change. Another source of variability is if the star has lot of spots like the sunspots. Again when the surface containing larger spotted area is facing the observer, the star's luminosity will be low. Likewise, bright regions on the surface come from close to the magnetic poles of the star. These will also add to variability.

Binary stars: We have already mentioned how binary stars such as the eclipsing binaries can show light-variability. Algol variable is one such example. Other examples of eclipses are Beta Lyrae variables which are extremely close binaries for whom it is difficult to resolve when the eclipse starts and when it ends. Stars in groups very tightly knit together such as in W Ursae Majoris, similarly present a variable appearance.

Gravitational waves

We end this chapter by once again stressing the possibility of gravitational waves providing us with additional aspects of stellar variability. A changing quadrupole moment theoretically produces gravitational waves. The intensity or power radiated is given by

$$L = \frac{G}{45c^5} \dddot{I}^2$$

where $\dddot{I}^2$ is the square of the third time derivative of the quadrupole moment. Asymmetrically exploding supernovae, magnetized spinning neutron stars, etc. are single sources for this radiation while coalescing binaries or merging black hole pairs going round each other are binary sources. At present several detectors are beginning operations of receiving such radiation. It is too early to say whether they will succeed or indicate the need for more sensitive versions.

Chapter 12

The Quest Continues

We end this account with the mention of a few emerging areas in this field, especially those which are expected to enlarge our horizons considerably.

Extra-solar Planets

While describing the overall picture of formation of a star in Chapter 5 we made it appear that just as a star is formed so are some planets to go round it. This picture naturally requires planets to exist around most stars. Do they? Till about 1991, the Sun was the only known star with planets. So the question remained unanswered. However, since then the situation changed dramatically so that today we can claim that extrasolar planets exist going around more than 500 stars! In fact the discovery of the first such planet came in a rather unexpected fashion.

In February 1990, Aleksander Wolszczan from the Aecibo Observatory discovered what turned out to be two planets going around a new pulsar PSR 1257+12. The discovery was made not by direct observations of the planets but by their gravitational perturbative effect on the pulsar, as shown by the very minute irregularities in the arrival times of pulses from the pulsar. Thanks to the precise time keeping of the pulsar, time changes as short as a fraction of a trillionth of a second could be detected and related to the periods of the planets. These studies led to the conclusion that the inner planet takes 66.6 days to go round the pulsar while the outer one takes 98.2 days. Theoreticians are now grappling with the problem of how planets came to be formed around a pulsar. For, as we have seen, the conventional wisdom suggests that planets form at the time of star formation, and not when the star has reached the end of its evolution to become a pulsar!

The first case of a planet orbiting a main sequence star like the Sun, came in 1995. A large planet named 51 Pegasi b was found going round the star 51 Pegasi. The method of detection is indirect in the sense that we do not see the planet but can measure the small perturbations of the star because of the gravitational field of the planet going round it. For the Earth-Sun type of system this type of effect is small because the Earth is not very massive and its location is not very near the Sun. In the case of 51 Pegasi, the planet is massive and is located near the star. Both these effects contribute to a vibratory motion of the star which gets reflected via the Doppler effect in the vibratory nature of its spectral lines. These can be measured with state of the art spectroscopes.

Another method, which requires somewhat special alignment, takes advantage of the transit phenomenon, when the planet is seen projected across the stellar disc, moving along its orbit. The star gets dimmed slightly when partly covered by the planet. This effect demands very sensitive photometry and can result in false identification. Thus, to be sure of a genuine case, a confirmation by some other method is advised.

Planets with large separation from their parent star can often be detected by the phenomenon of *gravitational microlensing*. Here the foreground (parent) star acts as a lens and bends the light rays from a background star and with suitable positioning amplify its brightness. This amplification is further modified by the planet going round the lensing star.

There are other methods which also help confirm the existence of a planet, but they may not be of universal use. While most data on extra-solar planets has so far come from ground based telescopes of modest apertures, there are two space missions currently searching for extrasolar planets. These are *COTOT* launched in December 2006 and *Kepler*, launched in March 2009. By February 2011, *Kepler* had identified 1,235 possible (but not confirmed) planetary candidates around 997 stars.

Searches for extrasolar planets or *exoplanets* are also motivated by the desire to find an answer to the question: *Are we alone in the universe?*. If direct or indirect methods tell us that the planetary environment is 'life-friendly' it brings us closer to an affirmative answer to the above question. For this reason attempts are made to find out the ecological conditions of the extra-solar planets.

The solar neutrino puzzle

We now come to a longstanding problem that relates to our Sun — a star we ought to know best. As we discussed in Chapter 6, the Sun is currently generating energy through the p–p chain in its interior. This chain produces a large flux of neutrinos. Neutrinos can escape from the deep interior of the Sun easily since they hardly take any notice of the matter around them. (Contrast this behaviour with that of photons who get very much knocked about before they eventually emerge from the Sun.)

Figure 12.1 illustrates the experiment set up deep underground in the Homestake mine by R. Davis to detect the neutrinos coming from the Sun. It consists of about 600 tonnes of CCl_4, which contains about 25% of ^{37}Cl as the detector. The weak process converts the chlorine to argon. The argon isotope that results is unstable. It is separated by flushing the system with ^4He. When it decays, an electron is produced, and this is detected. The half-life of this decay process is 35 days.

Neutrinos coming from the Sun arise from various processes, but those to which the above detection process is the most sensitive come from the decay of boron in an excited state.

(This is not, however, the reaction that generates the largest number of neutrinos in the Sun; that reaction is $p + p \rightarrow D + e^+ + \nu$, but the neutrinos generated by it are below the threshold energy for capture by the detector described above.)

One solar neutrino unit (SNU) corresponds to 10^{-36} captures per ^{37}Cl atom per second. Standard solar models predict count rates of about 7.8 SNU. The observed count rates are, however, 2.1 ± 0.3 SNU, far below the expected rate. Why this discrepancy? The discrepancy with only about one-fourth to one third of the predicted neutrinos

Figure 12.1: The neutrino detector experiment pioneered by R. Davis deep underground consists of a huge tank of fluid called perchlorethylene (C_2 Cl_4) that is exposed to the neutrinos coming from the Sun. The neutrinos interact with the chlorine nucleus in the solution and change it to argon which can be detected. Thus by measuring argon nuclei we estimate the neutrino flux. (*Photograph by courtesy of R. Davis, Jr., Brookhaven National Laboratory.*)

being detected, was serious enough to cause worry. Was the detector not functioning properly? Is the theory of the Sun's internal structure not quite correct? Has there been a gap in our understanding of nuclear reactions? Or is it our imperfect understanding of neutrinos that is the culprit?

All these alternatives were tried but no satisfactory explanation emerged. In the meantime, new generations of solar neutrino experiments started operating. Among these there was one at Kamiokande in Japan which looked for neutrinos scattered by electrons. The

Kamiokande II experiment has 680 tons of ultrapure water which acts as the detector. The results from this experiment yield about half the expected number of neutrinos from the Sun.

Another breed of detectors used gallium and were known by the acronyms SAGE and GALLEX. They started producing results from 1991–92. Here too the observed flux of solar neutrinos falls far short of the expected value, being in the range of 40%–60%.

The neutrinos being looked for in the different detectors fall in different energy ranges. All experiments have some statistical uncertainties associated with experimental errors. However, even allowing for these the discrepancy remained. Meanwhile another check on the standard solar model became possible.

In the 1980s, a useful probe of the solar interior came from the field of *helioseismology*. The subject arose in recent years from close studies of the disturbances of the solar surface. In fact, as early as during the 1960s periodic disturbances were noticed with a period of five minutes in patches covering half the solar surface. Known as "five minute oscillations" these turned out to be the tip of the iceberg! The Sun has seismic oscillations with periods much longer (20–60 minutes, 160 minutes, etc.)

The oscillations arise from internal fluctuations of the physical conditions of the Sun. Starting with a solar model one can deduce what type of oscillations one may expect to see, and then compare with what are seen. In this way we can check, change or confirm our postulates about the solar interior. Thus one finds that the spin found externally in the Sun's surface (like the Earth's spin about its polar axis) continues inwards but it does not increase rapidly with depth as some scientists had expected. Also, the abundance of helium inside is significantly high, higher than what theorists would like if they wish to solve the solar neutrino problem just described. Because of these dividends about our understanding of the Sun's interior, helioseismology has become an important field of study in solar physics.

Thus the feeling was growing in recent years that the apparent paucity of solar neutrinos came not from any defect of the standard solar model but had a possible origin in our understanding of neutrino physics. So the ball was firmly in the particle physicists' court. How well did they know the neutrinos?

Particle physics did offer a possible solution based on the assumption that neutrinos are not massless like the photons but possess small restmasses. This assumption violated the tenets of the standard model of particle physics which had hitherto worked well and which was centered on the assumption that neutrinos are massless. The three types of neutrinos, so far believed to exist, are the electron neutrino ν_e, the muon neutrino ν_μ and the tau neutrino ν_τ with progressively higher restmasses. If one does admit massive neutrinos then the theory does allow the phenomenon of *neutrino oscillations*. Under this phenomenon there can be spontaneous transformations of one neutrino species to another.

Given this possibility, suppose that of all the solar neutrinos created, a substantial fraction change over from the original electron neutrino to the other two types. These other types would not be detected by the detection systems which can pick up only the electron neutrinos.

The first direct evidence of solar neutrino oscillation came in 2001 from the Sudbury Neutrino Observatory (SNO) in Canada. It detected all types of neutrinos coming from the Sun, and was able to distinguish between electron-neutrinos and the other two flavors. After extensive statistical analysis, it was found that about 35% of the arriving solar neutrinos are electron-neutrinos, with the others being muon- or tau-neutrinos. The total number of detected neutrinos thus agrees quite well with the earlier predictions from nuclear physics, based on the fusion reactions inside the Sun. Although a lot still needs to be done in particle physics to improve our understanding of neutrinos, including the determinations of their restmasses, so far as the Sun is concerned, it is now assumed that the longstanding puzzle of solar neutrinos is finally resolved.

Gamma ray bursts

As evident from our discussion of the supernova phenomenon, the core collapse and the ejection of the envelope pose the most difficult challenges to a mathematical description of what goes on in the star. Thanks to the availability of fast computers with vast memory space, time dependent partial differential equations relating to this description can now be tackled. Considerable progress is expected on this front in the years to come.

However, in recent years a phenomenon on a grander scale than the typical supernova has been discovered, thanks to the growth of gamma ray astronomy. Known as *Gamma Ray Burst*, in short 'GRB', this was first discovered in the 1960s, the first sighting coming on July 2, 1967. The observing facility, the *Vela* satellites did not belong to the astronomers but to the military. The main purpose of these satellites was to monitor the operation of the Nuclear Test Ban Treaty. The GRBs as the name implies are brief flashes of intense gamma ray emission. Over the following thirty years hundreds of GRBs were detected but technology did not permit their detailed study. Lack of accurate imaging meant that these flashes came and went without, apparently, leaving any 'aftermath'.

The Burst and Transient Source Experiment (BATSE) launched on board the *Compton Gamma ray Observatory* in April 1991, measured about 3000 events, thus suggesting that on a typical day two to three GRBs occur all over the observable universe. A typical burst lasts a few seconds only although it can outshine every other source in the gamma ray sky. Broadly, the sources are classified as short or long depending on their lifetime being less or more than two seconds. Spectroscopically, the two types differ: the short bursts release relatively more high energy γ-rays than the long ones.

While the early measurements were limited, a sea-change in the situation occurred in 1997 when the *BeppoSAX* satellite obtained high resolution X-ray images of the predicted afterglow of the source GRB970228 (so named as to indicate that it was seen as a burst on February 28, 1997). This allowed precision in locating the source and its optical identification as well as further investigation in that waveband. Further investigation could tell, for example that the source was extragalactic and located in a galaxy whose redshift was measurable. More GRBs could be studied similarly. Indeed, the afterglow of a GRB is now routinely observed and these observations will eventually help us in solving the mystery of what a GRB is.

What do we know about a GRB? If we estimate the energy emitted by a burst in its initial phase it is as high as $10^{53}\Omega/4\pi$ erg, provided we take its distance as estimated by Hubble's law using the redshift of its host galaxy. Here Ω is the solid angle over which the emission took place. The seat of activity is very compact with linear size less than 300 km. That is, in the isotropic case we are expecting the energy equivalent of $0.06M_\odot$ emitted from such a compact region over a couple of seconds. Since the energy emitted by the GRB in gamma rays is derived from some basic source with an efficiency factor $\epsilon < 1$ the requirement of the basic source is further increased by the reciprocal of this factor. To ease the tight energy requirement it is assumed that relativistic beaming is going on with the source expanding with velocity very close to the speed of light. It is believed that a large supernova which is exploding relativistically in a non-isotropic fashion can foot the bill! The 'non-isotropy' implies a jet like ejection directed almost towards the observer. The source appears with enhanced brightness if seen by such an observer. Astrophysical arguments can be given to show that the spectrum of emission will soften with time as is observed.

Perhaps it is worth mentioning here that in 1974, in a paper in *Nature*, K.M.V. Appa Rao, Naresh Dadhich and the author had proposed a *white hole* as a model for such bursts. The time-reversed version of a black hole, the white hole is essentially an exploding object that emerges out of a very compact volume with relativistic speed slowing down with time. The emission from such a source is strongly blueshifted to start with but the blueshift reduces with time as the exploding energy is reduced. This leads to a softening of the spectrum.

No doubt with further observations of GRBs close after their occurrence we will find the basic cause of this remarkable phenomenon. The supernova hypothesis, if borne out will show the GRB to be the most dramatic event in the stellar community. If the white hole version stands up to the observational tests then that will show the role white holes can possibly play in other contexts in phenomena involving high energy astrophysics.

Problems in nuclear astrophysics

As we have seen, nuclear astrophysics plays a crucial part in the studies of stellar evolution and nucleosynthesis. We now briefly describe areas where further research will bring greater enlightenment.

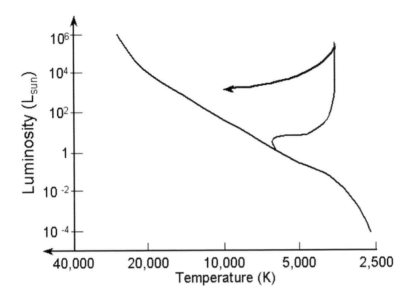

Figure 12.2: A detailed understanding of the so called horizontal branch stars is still awaited as a link between the giants and the dwarfs.

The evolution of less massive stars from red giant to the white dwarf state is not yet worked out in quantitative detail. Here again ingenious use of computing techniques in rapidly changing physical situations will ultimately deliver the answer. The so-called *horizontal branch* stars fill the gap between the giants and the dwarfs as seen in Figure 12.2. Clearly the stars must make a continuous transition along such a route. While gravity will play the key role what is the status of nuclear reactions towards the end of star's nuclear fuel?

The mass limit on stable neutron stars is another problem of current and future research, for it has bearings on the question of how massive a star has to be so that it ultimately becomes a black hole. This subject takes us to studies of nuclear matter at densities exceeding a million billion times the density of water. Do other forms of highly dense matter exist which could support a star as massive as $5M_\odot$ or $10M_\odot$ towards the end of its active nuclear life? One needs also to answer if there are further stages beyond the neutron star, like the *hyperon star* or the *quark star* signifying even denser stages.

The calculations of stellar evolution and nucleosynthesis estimate the age of the star systems we observe. For example, we have already seen that a star spends the longest period of its active life on the main sequence. A system of stars that has completed its main sequence phase and branched off to the giant phase will therefore be quite old. Globular clusters are such star systems of which we saw an example in Figure 7.6. Knowing the vital statistics of these stars and the rate at which the main sequence nuclear fusion proceeds we can estimate the ages of globular clusters. Conservative estimates of these ages are about 12 billion years, although values as high as 18 billion years cannot be ruled out. Stars less massive than the Sun, say, of $0.7M_\odot$ in a similar state would be as old as 20 billion years. Do such stars exist?

A second "aeon glass" for timing stellar events is provided by the radioactive decays of elements produced in the r-process of nucleosynthesis (briefly referred to in Chapter 7). For example thorium (^{232}Th) decays on a radioactive time scale of 20 billion years while the uranium isotopes ^{238}U and ^{235}U have time scales of 6.51 and 1.03 billion years respectively. Thus the ratio of abundances of these elements, ^{232}Th/^{238}U and ^{235}U/^{238}U in meteorites can give us estimates of the time when these elements were first manufactured in stars. The decay of rhenium (^{187}Re) to osmium (^{187}Os) is another example of long term radioactivity. The time estimates again turn out to be as high as 12–20 billion years.

There is a problem posed by these apparently long ages to cosmologists who study the structure and evolution of the universe as a whole. The currently favoured picture of the origin of the universe suggests that it exploded into existence in a gigantic explosion, the *big bang*. Estimates of how long ago the big bang took place depend on measurements of how fast the universe is expanding *at present*. The time scale indicated by present observations puts the big bang epoch 13.8 billion years ago. It is therefore very important that the study of stellar evolution provides us with independent lower limits on the age of the universe.

The second count on which the information will be useful is in deciding to what extent life might exist elsewhere in the Galaxy. A planet located at a suitable distance from the parent star (as the Earth is from the Sun) and with suitable nuclear composition, may be in a position to host life and support it with the energy received from the star.

If planetary systems are detected in a few of such neighbouring stars, astronomers may direct some of their big radio telescopes towards intercepting intelligent information from the supposedly advanced civilizations located on them!

Needless to say, a positive result of such explorations will add an extra dimension of excitement to astronomy.

Epilogue

Its mission accomplished the spaceship made its way out of the Solar System. The Professor who was reading the detailed report prepared by Sunya began to smile. Shortly he beckoned Sunya. Expecting another reprimand, Sunya was pleasantly surprised to see his mentor in a good mood. The Professor said: "Do you know Sunya, that by sheer fluke you have picked up an exceptional human being that no amount of Purna's statistics can help us with? Your trip was not wasted."

The picture of Sunya's human case study is shown in the adjoining figure.

While relying mainly on statistical surveys of heavenly bodies the astronomers are also on the lookout for the rare and the spectacular like the Crab Nebula and Cygnus X-1.

The human being studied by Sunya.

Appendix: Powers of Ten, Logarithms and the Magnitude Scale

Astronomers find powers of ten convenient to describe large numbers. The power notation is simple to understand and will be clear to the uninitiated reader from the following simple examples:

$$10^2 = 100, \quad 10^4 = 10,000, \quad 10^6 = 1,000,000.$$

The power, that is the number appearing on the top right above 10 seems from the above example, to be just the *number* of zeros after one. In fact, an alternative and more precise way of defining the power of ten is through the *number* of factors 10 in the concerned expression; thus

$$
\begin{aligned}
10^2 &= 10 \times 10, \\
10^4 &= 10 \times 10 \times 10 \times 10, \\
10^6 &= 10 \times 10 \times 10 \times 10 \times 10 \times 10.
\end{aligned}
$$

When such numbers multiply their powers add simply because the numbers of constituent factors are added:

$$10^2 \times 10^4 = (10 \times 10) \times (10 \times 10 \times 10 \times 10) = 10^6 = 10^{2+4}.$$

Mathematically speaking the opposite operation of raising the power of ten is that of taking the logarithm. Thus the question "What is the number obtained by raising 10 to the power 6?" has the answer "One million." Turn the original question round to: "What power must 10 be raised to in order to get a million for the result?" and the answer must be 6. This statement is written as

"logarithm of a million to base ten is six",

or in a compact notation

$$\log_{10}(1,000,000) = 6.$$

It is usual to drop the explicit mention of "base 10" from this notation with the understanding that we are talking about powers of ten only. Thus

$$\log(1,000,000) = 6, \quad \log(10,000) = 4, \quad \log(100) = 2.$$

It is obvious that as we raise 10 to higher and higher powers we get larger and larger numbers. Conversely, the logarithm of a number increases as we increase the number.

This raises the question: "Can we assign logarithms to numbers that are not obtained by simply multiplying 10 any number of times?" The answer is yes, if we make use of the above property.

Thus we know that

$$\log 100 = 2, \qquad \log 1000 = 3.$$

What is the logarithm of a number lying between 100 and 1000? By the above property it must lie between 2 and 3. Somewhat advanced mathematics is needed to compute the actual answer. The task is, however, simplified in practical terms by tables of logarithms. These tables give ready-computed values of logarithms of such intermediate numbers to any desired degree of accuracy. Thus from four figure tables we know that

$$4\log 200 = 2.3010, \qquad \log 300 = 2.4771.$$

As expected, the answer lies between 2 and 3; also the logarithm of 300 is greater than the logarithm of 200. Tables of logarithms exist from which one can read off the logarithm of any number and vice versa.

With the help of logarithms we can look upon any positive number as a power of ten; only the power may not be a whole number:

$$200 = 10^{2.3010}, \quad 300 = 10^{2.4771}.$$

Corresponding to the addition of powers under multiplication, we get the law of addition of logarithms. Thus the product

$$100 \times 10,000 = 1,000,000$$

$$\log(100) + \log(10,000) = \log(1,000,000).$$

Let us apply this law to the magnitude scale of Chapter 3. We know that a difference of five magnitudes represents a factor of 100 in luminosity. What ratio does a difference of one magnitude corresponds to? Suppose the answer is R. Then the law of addition of logarithms tells us that because,

$$R \times R \times R \times R \times R = 100$$

we have

$$5 \log R = 2.$$

Therefore, $\log R = 2/5$. From log tables we find that R is approximately equal to 2.512.

Credits for Images

2.6 : Indian Institute of Astrophysics, Bangalore

4.1 : David Malin/Anglo Australian Observatory

4.5 : From *On the Construction of the Heavens* by William Herschel, published in Philosophical Transactions of the Royal Society of London, Vol. 75 (1785), pp. 213–266

4.11 : NASA

4.13 : NASA

5.1 : Observatories of the Carnegie Institution of Washington

5.2 : National Radio Astronomy Observatory, operated by Associated Universities

8.1 : William C. Miller

8.2 : NASA

8.4 : David Malin/Anglo Australian Observatory

8.5 : NASA

8.6 : NASA

9.6 : Mullard Radio Astronomy Observatory, Cavendish Laboratory, Cambridge

10.4 : NASA

12.1 : R. Davis Jr., Brookhaven National Laboratory

Index